APHID ECOLOGY

To Dick Hille Ris Lambers and George Varley

APHID ECOLOGY

A. F. G. DIXON, B.Sc., D.Phil.

Professor of Biological Sciences
University of East Anglia

Blackie

Glasgow and London
Distributed in the U.S.A. by
Chapman and Hall
New York

Blackie & Son Limited
Bishopbriggs
Glasgow G64 2NZ

Furnival House
14–18 High Holborn
London WC1V 6BX

Distributed in the USA by
Chapman and Hall
in association with Methuen, Inc.
733 Third Avenue, New York, N.Y. 10017

First published 1985

British Library Cataloguing in Publication Data

Dixon, A. F. G
Aphid ecology
1. Plant-lice
I. Title
595.7′52 QL527.A64
ISBN 0–216–91647–X

For the USA, International Standard Book Number is
0–412–00381–3

Photoset by Thomson Press (India) Limited, New Delhi
and printed in Great Britain by McCorquodale (Scotland) Ltd

Preface

This book is about the ecology of a group of small, sap-sucking insects, the aphids, some of which are serious pests of agricultural crops and forest trees. The association of aphids with plants is considered in relation to certain of the features that aphids developed early in their evolution. In particular, the book explains how the modes of feeding and reproduction of aphids have led to a very close and specific association with plants and have affected their size and population structure, and it accounts for the paucity of species in the tropics.

This book is written for specialists, postgraduates and advanced undergraduates in entomology, but I hope that other people will find it interesting.

I am grateful for the opportunity to acknowledge the help of all my students, in particular Seamus Ward, in developing the ideas presented here. A number of people have helped in the preparation of this book. I am especially grateful to my wife, to Angela Risebrow and to Julian Entwistle for reviewing the manuscript and offering plentiful and excellent advice, and to Drs Aoki, Bromley, Gibson, Jukes and Tjallingii for supplying photographs used in the text. I also wish to thank Sue Mitchell for preparing the figures, and Gwen Thompson for typing the manuscript.

A.F.G.D.

Contents

1 Introduction

Aphids have fascinated and frustrated man for a very long time. This is mainly because of their intricate life style in close association with their host plants, their polymorphism and ability to reproduce both asexually and sexually. Thus aphids are ideal for studying many of the topical issues in ecology, the term 'ecology' being used in a broad sense to include aspects of the basic biology of the group necessary for an understanding of the population and community levels of organization. The modes of feeding and reproduction that developed early in aphid evolutionary history led to a close and specific association with certain plants and affected aphid size and population structure. In turn, these consequences have influenced aphid life history patterns, dispersal behaviour, population dynamics, species diversity and world-wide distribution.

There are about 4 000 described species of aphid. Compared with 10 000 grasshoppers, 12 000 geometrid moths and 60 000 weevils, the aphids are a contained group. However, their diversity is expressed as polymorphism as well as speciation. The greatest number of aphid species occurs in the temperate regions, and there one plant species in four is infested. However, the relatively few species of aphids among insects is in itself intriguing, because like so many other aspects of aphid ecology, the explanation is to be found in the basic biology of the group.

Unfortunately there are few biological studies on aphids indigenous to Africa, South America and India. Until such gaps in our knowledge are filled, an account of aphid ecology (like the present one) will be unbalanced because it is based mainly on our information about temperate-region aphids. The results of future studies of aphid biology in the equatorial and the hot dry regions of the world are likely to challenge, as well as complement, the generalizations about aphid biology that are founded on the studies of these insects in temperate and markedly seasonal environments.

Distinguishing features of aphids

They are small (1–10 mm), soft-bodied, plant-sucking insects. Several or all generations comprise parthenogenetic females which do not require fertili-

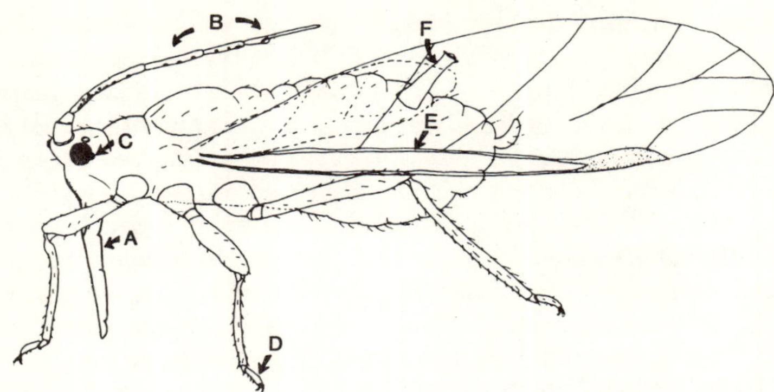

Figure 1.1 Diagnostic morphological features of aphids. (After Heie, 1980.)

zation and are viviparous. Species in which periods of asexual reproduction alternate with sexual reproduction are said to show cyclical parthenogenesis. As the eggs of parthenogenetic females commence development immediately after ovulation, a nymph can have embryos developing within itself that also have embryos. This telescoping of generations and parthenogenesis enable aphids to achieve very high rates of increase. The winged aphids are known as alatae, and wingless aphids, as apterae. Polymorphism, which is the occurrence within a species of different forms or morphs, is also a characteristic of aphids.

The more obvious diagnostic morphological features are illustrated in Figure 1.1. A: the base of the proboscis lies between and behind the fore coxae; B: the antennae have two short thick basal segments and a thinner flagellum, of at most four segments, the ultimate one of which consists of a proximal part and a thinner distal part, the process terminalis; C: there is an ocular tubercle made up of 3 lenses (a triommatidium) situated behind each compound eye; D: there are two tarsal segments; E: the wings have only one prominent longitudinal vein, and F: there is a pair of siphunculi on the dorsum of the fifth abdominal segment.

Origin of aphids

Aphids, adelgids and phylloxerids belong to the insect superfamily Aphidoidea, within the order Homoptera, the plant-sucking bugs. They are all small insects; wings, if present, are membranous. Although the oldest fossil aphid known, *Triassoaphis cubitus*, is from the Triassic, it is likely that the Aphidoidea evolved 280 million years ago, in the Carboniferous, from the same stock that gave rise to the now extinct Archescytinidae and Permaphidopsidae, which are thought to have lived on primitive gymnosperms like the Cordaitales and Cycadophyta. Reproduction by means of unfertilized eggs is common to all three families of the Aphidoidea, and as the adelgids and phylloxerids appeared in the late Carboniferous or early Permian, parthenogenesis possibly evolved before these three families separated, over 200 million years ago. The resultant cyclical parthenogenesis, in which periods of parthenogenetic reproduction alternate with sexual reproduction to give a holocycle, is thought to have evolved in a seasonal climate, possibly that associated with the glacial period in the Lower Permian. However, viviparity, another characteristic feature of aphids, must have evolved later, as the Archescytinidae and the modern adelgids and phylloxerids are oviparous (Heie, 1967). The characteristic shape and veining of their wings, and the structure of their proboscis and legs, had evolved by the Jurassic, whereas the cauda and siphunculi appeared later in the Cretaceous (Shaposhnikov, 1977).

Primeval aphids were probably polyphagous and fed on the parenchymatous and phloem tissues of plants. That the Aphidoidea are a quarter of the size of the Archescytinidae and Permaphidopsidae, Heie (1967) attributes to their parasitic mode of life and their use of air currents for dispersal. He sees monophagy as a recent development in aphid evolution. However, as most of the extant species of all three families are monophagous it could be argued that monophagy, like parthenogenesis, evolved early in the history of the Aphidoidea. The simple nymphal eye of 3 lenses, the triommatidium, is a characteristic which aphids share with adelgids and phylloxerids and is also associated with the reduction in body size that came with the evolution of a parasitic way of life.

The Aphidinae, the largest subfamily of modern aphids, is not represented in the fossil record until the late Tertiary and is associated with the evolution of the angiosperms, in particular the Rosaceae. Tribes of the second largest subfamily, the Drepanosiphinae, developed much earlier in the Upper Cretaceous and early Tertiary. There is no palaeontological evidence of the evolution of the third largest subfamily, the Lachninae. As eighty percent of them live on conifers, which are evolutionarily older than the angiosperms, the lachnids have generally been regarded as primitive. However, two genera, *Aphis* in the Aphidinae and *Cinara* in the Lachninae, each have a large number of species that are difficult to distinguish from one another. This and the absence of fossil lachnids led Heie (1967) to suggest that the Lachninae are of recent origin.

Classification

Börner (1952), Shaposhnikov (1964), Eastop (1977), and Heie (1980) have classified the Aphididae. The scheme presented in Table 1.1 is that of Eastop. In general, there is little disagreement about the number or delimitation of subfamilies. However, whether the Lachninae should be regarded as one of the most primitive or one of the most advanced groups of aphids is a major issue on which there is no consensus.

Table 1.1 Family Aphididae

Subfamily	*Tribes*	*Percentage of total species in each subfamily*
1. Lachninae	Cinarini Lachnini Tramini	9
2. Chaitophorinae	Chaitophorini Siphini	4
3. Drepanosiphinae	Phyllaphidini Drepanosiphini Neuquenaphidini Lizeriini Mindarini Neophyllaphidini Thelaxini Saltusaphidini	12
4. Pterocommatinae		1
5. Aphidinae	Aphidini Macrosiphini	59
6. Greenideinae	Greenideini Cervaphidini	3
7. Phloeomyzinae		(1 species only)
8. Anoecinae		1
9. Hormaphidinae	Hormaphidini Nipponaphidini Cerataphidini	4
10. Pemphiginae	Eriosomatini Pemphigini Fordini	7

Ease of identification largely determines the attractiveness of insects for study. As individuals of most species of aphids exist in one of several different forms, because they are polymorphic, they are difficult to identify. However, modern keys to aphids exist or are being developed for most of Northern Europe (Heie, 1980, 1982; Hille Ris Lambers, 1938, 1939, 1947, 1949, 1953; Shaposhnikov, 1964; Stroyan, 1977), North America (Palmer, 1952; Richards, 1960, 1963, 1965, 1972), the Middle East (Bodenheimer and Swirski, 1957), East and West Africa (Eastop, 1958, 1961), India (Ghosh, 1980, 1982), Japan (Shinji, 1941; Miyazaki, 1971; Higuchi, 1972; Aoki, 1975; Akimoto, 1983), Korea (Paik, 1965, 1972), Australia (Eastop, 1966) and New Zealand (Cottier, 1953).

Distribution

Although predominant in temperate regions, aphids have a world-wide distribution. Within the Aphididae, the Aphidinae and Drepanosiphinae, that make up 70% of modern aphids, are not restricted to a particular region, but the Greenideinae and Homaphidinae, that make up 7%, are restricted mostly to South East Asia and Australia, although the presence of fossil Greenideinae in Yugoslavia indicates that this group once had a much wider distribution (Eastop, 1977). However, as the most important families of aphids are not restricted in their distribution we have to account for why there are so few species of aphids in the tropics, and this is considered in Chapter 9.

Summarizing, aphids evolved 280 million years ago, and from the beginning showed two of the more striking features of modern aphids: small size and parthenogenetic reproduction. They have a world-wide distribution and although certain species can become very abundant, there are surprisingly few species, especially in the tropics. Polymorphism and the telescoping of generations are also characteristic of aphids.

2 Host selection

Plants are colonized primarily by flying aphids that, because of their low speed, have little control over the direction of their flight. However, once within the layer of relatively still air around vegetation, called the 'boundary layer' by Taylor (p. 99), aphids can control their landing on plants and respond to either olfactory or visual cues, or both. After settling, an aphid recognizes a potential host by the structure and chemistry of its surface and internal tissues. The distribution of aphids between and on the host plant(s) is largely determined by variations in the quality of phloem sap. Plant species recognition and quality form the basis of the dual discrimination theory of host selection (Kennedy and Booth, 1951).

Settling behaviour

Aphids appear to respond predominantly to visual cues, showing a marked orientation to yellow. Colour is a good indicator of the nutritive status of a plant, as both highly nutritious young and senescent foliage tend to be yellower than the nutritionally poorer mature leaves, but a poor indicator of a plant's taxonomic status (Kennedy *et al.*, 1961). In spite of the distinctive odour of many plants there is no unequivocal evidence (Pettersson, 1970; Chapman *et al.*, 1981) that aphids are attracted by odour when locating their host plants. Carvone-baited water traps collect more carrot aphids (*Cavariella aegopodii*) than unbaited traps. However, the monoterpene carvone is not restricted to the Umbelliferae, the plant family to which the secondary host plants of the carrot aphid belong, and at least one of its favoured secondary host plants, *Anthriscus sativus*, lacks the substance. Therefore, although this response provides evidence of odour-induced orientation behaviour in aphids, carvone is clearly not the only factor used in host finding by *C. aegopodii* (Chapman *et al.*, 1981).

Thus the accumulation of aphids on their host plants is mainly due to a differential rate of departure, with aphids staying longer on their host plants (Kennedy *et al.*, 1959). The rate of departure is determined by the aphid's response to contact with the external and internal features of a plant.

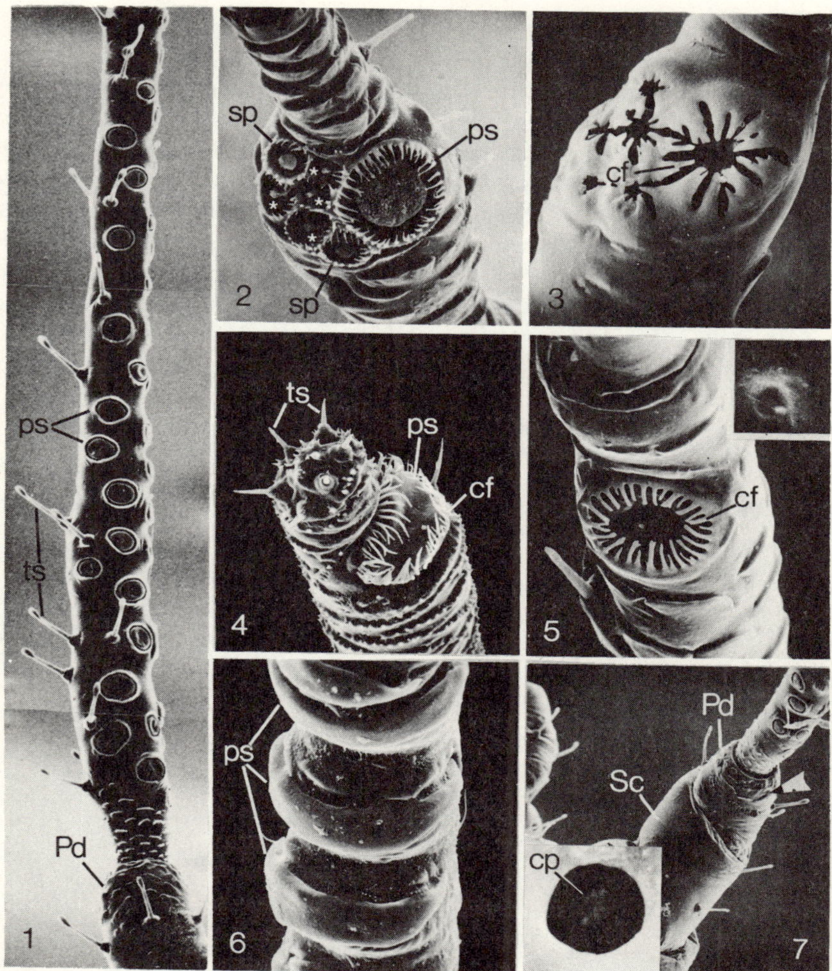

Figure 2.1 Scanning electron micrographs of aphid antennal sensilla. 1. Small placoid sensilla (secondary rhinaria) on third segment of alate *Nasonovia ribis-nigri*, × 200; 2. Primary rhinarium on sixth segment of alate *Aphis pomi* comprising one large placoid sensillum, two smaller ones and four coeloconic pegs, × 900; 3. Primary rhinarium on sixth segment of alate *Nasonovia ribis-nigri* almost totally enclosed by cuticular fringe, × 1000; 4. Primary rhinarium on sixth segment of alate *Pemphigus bursarius* (sexuparae) comprising a single enlarged placoid sensillum and four coeloconic pegs (not visible), × 680; 5. Primary rhinarium on fifth segment of *Macrosiphum euphorbiae*, × 1000. Inset: raised projection on outer cuticle of sensillum, × 9900; 6. Elongate placoid sensilla (secondary rhinaria) on third segment of alate *Pemphigus bursarius* (sexuparae), × 800; 7. Pedicel of alate *Nasonovia ribis-nigri* showing position of coeloconic sensillum (arrow), × 150. Inset: top of coeloconic sensillum at entrance to a pit, × 6750. (cf—cuticular fringe; cp—cuticular projections; Pd—pedicel; ps—placoid sensillum; Sc—scape; Sp—small placoid sensillum; ts—trichoid sensillum). (Reproduced with permission from Bromley *et al.*, 1979).

Host recognition

After alighting on a plant aphids walk over the surface testing it with their antennae and by probing it with their mouthparts. The antennae have many sensilla (Figure 2.1), amongst which there are some whose structure and electrophysiological responses indicate that they are used in close-range chemoreception or gustation and the perception of the leaf surface (Shambaugh *et al.*, 1978; Bromley *et al.*, 1979; 1980; Bromley and Anderson, 1983). Sensilla at the tip of the proboscis (Figure 2.2) are of two types: tactile receptors and chemoreceptors. Tactile receptors respond to contact and surface texture and enable aphids to detect the contours of veins, the preferred feeding site of many aphids (Tjallingii, 1978). The chemoreceptors described by Tarn and Adams (1982) probably detect the product of the interaction

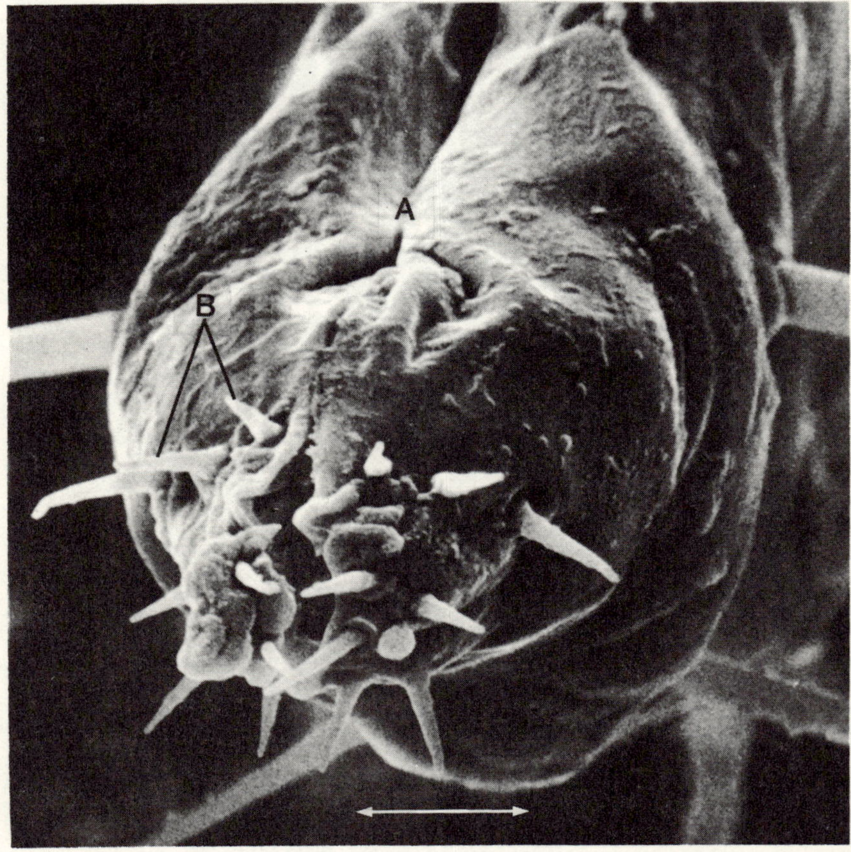

Figure 2.2 Scanning electron micrograph of the tip of the proboscis of *Myzus persicae* showing the labial groove (*A*) and 16 bilaterally symmetrically arranged sensilla (*B*). Scale bar is 15μm. (Photographed by Dr. W. F. Tjallingii).

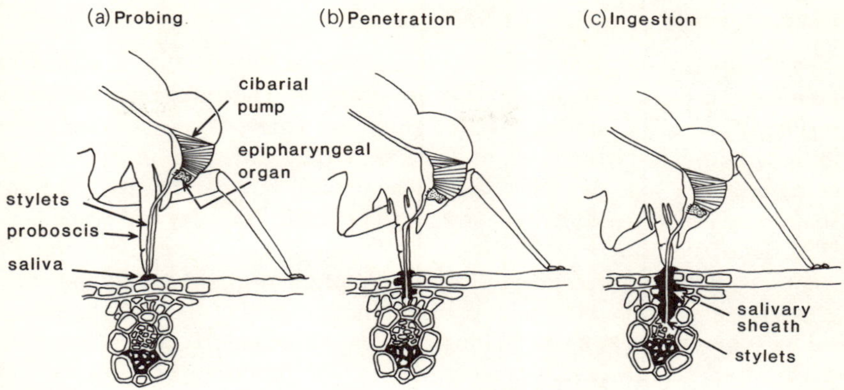

Figure 2.3 Diagram of the probing, penetration and ingestion phases of host selection.

between aphid saliva and plant epidermis. Although the mandibular stylets contain nerves (Forbes, 1966; Parrish, 1967) there is neither morphological nor unequivocal electrophysiological evidence for chemoreceptors at the tips of the stylets. However, in the dorsal wall of the food canal, at the base of the stylets and anterior to the dilator muscle of the cibarial pump, there is a group of sense endings called the epipharyngeal organ, that has the typical structure of a contact chemoreceptor (Wensler and Filshie, 1969). This organ is in direct contact with fluid in the food canal and possibly detects chemicals in the ingested sap (Figure 2.3).

While walking over and probing the surface of a plant, an aphid obtains information about the physical properties and chemistry of the surface and the internal chemistry of the plant. This initial investigation of the surface involves little or no stylet penetration, but often enables an aphid to sense the suitability of a plant within 60 seconds. This is the time it takes an aphid to penetrate the epidermis and exude saliva on the surface (Harris, 1977; Pollard, 1977) (Figure 2.3*a*). Once an aphid settles it probes deep into the plant (Figure 2.3*b, c*).

While probing plant tissues aphids secrete saliva that forms a sheath around the stylets. The salivary sheath gives rigidity to the very flexible stylets and enables aphids to control the direction of the probe by restricting the bending except at the apex of the stylets (Pollard, 1973). The stylet sheath usually ends in the phloem, indicating that aphids feed on the sieve tubes (Pollard, 1973). Phloem sap is under 15–30 atmospheres of pressure, sufficient to force sap through the extremely fine food canal in an aphid's stylets, and one could suppose that aphids feed passively. However, anaesthetized aphids not only cease to excrete but neither do they leak or swell up (Mittler, 1957). It is likely that they use their cibarial pump (Figure 2.3) to control the flow of sap (Banks, 1965) and are even able to suck up sap, as when feeding on synthetic diets.

It takes approximately 40 minutes for the black bean aphid (*Aphis fabae*) to

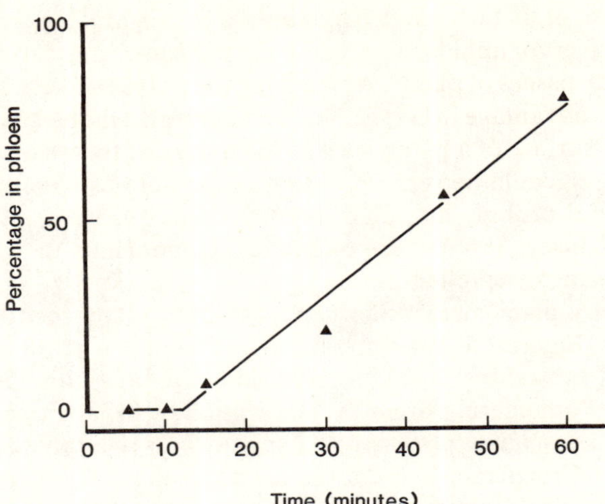

Figure 2.4 Percentage of *Aphis fabae* with stylets inserted in the phloem in relation to the time from start of penetration. (After Henning, 1966.)

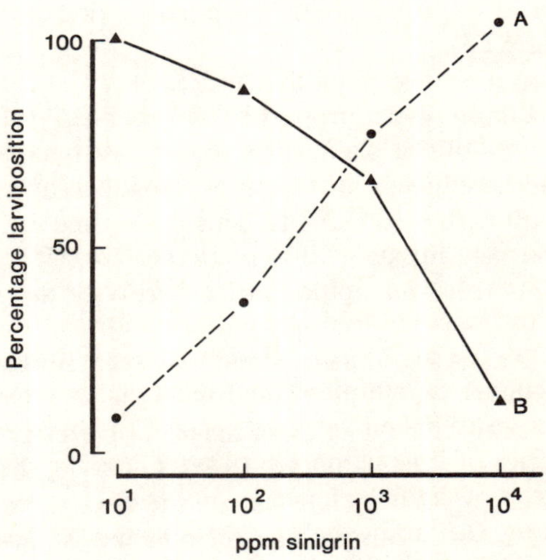

Figure 2.5 Percentage of *Brevicoryne brassicae* and *Acyrthosiphon pisum* producing offspring on cuttings of broad bean in relation to the concentration of the sinigrin solution supplied to the cuttings. Broad bean is a good host for *A. pisum.* (After Nault and Styer, 1972.)

reach the phloem of its host plant (Henning, 1966; Figure 2.4), and possibly considerably longer for aphids that feed on the phloem elements situated deep within the woody tissues of plants. As reaching the phloem takes so long there is a considerable advantage in being able to recognize a suitable plant quickly. Features on the surface of a plant, such as the nature of the waxes (Klingauf, 1972) or hairiness, could be very important in the initial stages of selection. During superficial probing an aphid imbibes small quantities of plant sap (McLean and Kinsey, 1968) specific chemical components of which could stimulate the gustatory epipharyngeal organ. Many are known to respond to specific secondary plant metabolites. This has been established by making non-host plants (Figure 2.5) and synthetic diets acceptable by the addition of secondary plant substances. The glucoside phlorizin, present in the leaves of apple and other Pomoideae, promotes colonization by the apple aphids *Aphis pomi* and *Rhopalosiphum insertum* (Klingauf, 1971). Similarly, sinigrin, characteristic of Cruciferae, induces the cabbage aphid (*Brevicoryne brassicae*) to settle and feed (Wensler, 1962; Nault and Styer, 1972); and rutin and quercetin have the same effect on the dock aphid (*Aphis rumicis*) (Herger, 1975). It is interesting to note that these structures and substances often act as barriers to the colonization of plants by many pathogens and herbivores, but enable host specific aphids to recognize their host plants. These devices also protect plants against colonization by polyphagous species of aphids and are a barrier to the extension of host range in host-specific aphids.

The physical defences of plants

The leaves and stems of many species of plants are covered with small epidermal hairs and/or hooks. The function of this pubescence has rarely been investigated and it is unknown whether host-specific aphids respond to the pubescence in selecting their host plants.

The undersides of the leaves of limes like *Tilia petiolaris* appear white because of the dense covering of stellate hairs (Figure 2.6). Adult lime aphids are able to feed on the leaves of this lime because there are few hairs on their preferred feeding site, the main veins. However, newly-born nymphs starve because the dense barrier of interdigitating stellate hairs prevents them from reaching the smaller veins. When the pubescence is removed young aphids do as well on this species of lime as on their native host, *T. platyphyllus*. The underside of the leaves of the silver lime (*T. tomentosa*) is covered with stellate hairs 25 to 35 μm long. In Turkey the leaves of this species provide food for a species of lime aphid whose proboscis is 20 to 40 μm longer than that of the European lime aphid: the specialist aphid has overcome the barrier by virtue of its longer reach (Carter, 1982).

The foliage of the wild potato (*Solanum berthaultii*) is covered with two types of glandular hairs. Type A has a short stalk and a four-lobed head that ruptures on contact and releases a quick-setting fluid, and Type B has a longer

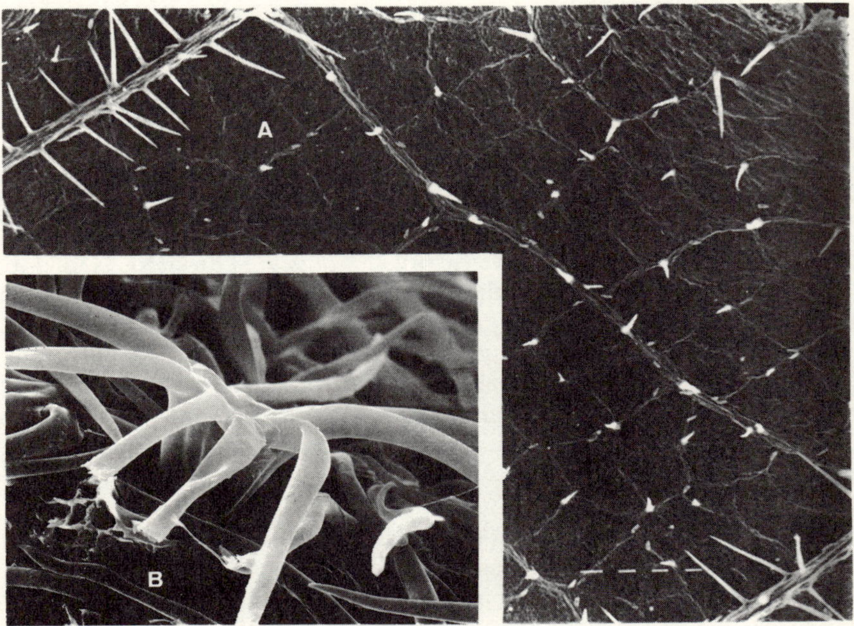

Figure 2.6 Scanning electron micrographs showing the simple hairs positioned along the leaf veins of (*A*) *Tilia platyphyllos* and a stellate hair on a leaf of (*B*) *T. petiolaris*. (Photographed by M. Jukes.)

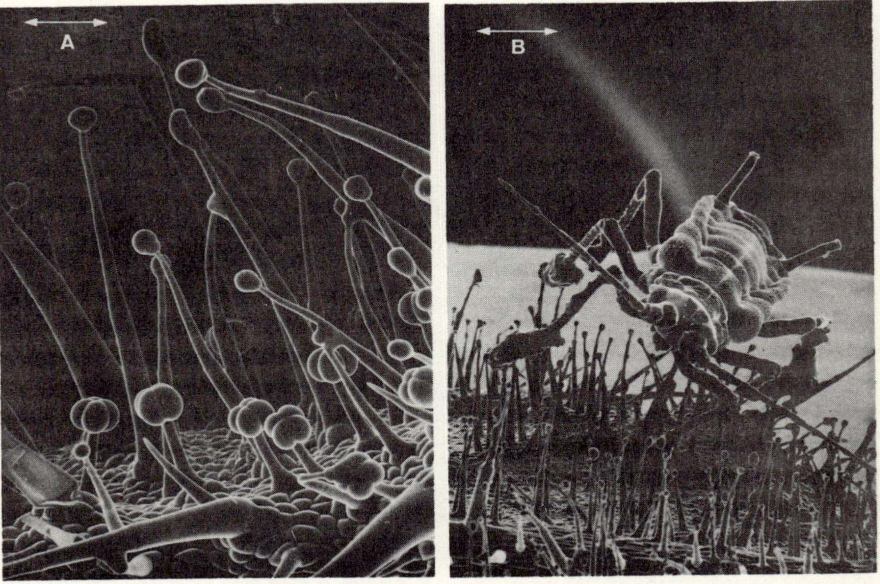

Figure 2.7 Scanning electron micrograph of the two types of glandular hairs on the leaves of *Solanum berthaulti* (*A*) and of an aphid entangled in the secretion of the glandular hairs (*B*). Scale bar is 97 μm in *A* and 290 μm in *B*. (Photographed by Dr. R. W. Gibson.)

stalk and exudes a sticky droplet at the tip (Figure 2.7*A*). The exudate of the Type B hairs supposedly entangles aphids that then struggle and rupture Type A glands, whose contents physically trap the aphids (Figure 2.7*B*) (Tingey and Laubengayer, 1981; Gibson and Pickett, 1983). It is the high density of these glandular hairs on the foliage, and not the presence of high levels of glycoalkaloids in the tissues of species of wild potatoes, that accounts for their resistance to aphids (Tingey and Sinden, 1982).

The chemical defences of plants

In addition to the physical defence afforded by glandular hairs of plants, the hairs may also produce and release insecticides that are highly effective against aphids. For example, nicotine, nornicotine, and anabasine are produced by the glandular hairs of species of tobacco (*Nicotiana*) and *Petunia* (Thurston *et al.*, 1966; Thurston, 1970) and 2-trideconone by those of the wild tomato (*Lycopersicon hirsutum f. glabratum*) (Williams *et al.*, 1980). The Type B glandular hairs of potato even release the aphid alarm pheromone, (*E*)-*β*-farnesene, and maintain a concentration around the foliage sufficient to result in an alarm response in aphids and make them avoid the foliage (Gibson and Pickett, 1983).

Aphids that reach the phloem of resistant cultivars of their host plant tend to cease feeding shortly after the phloem is penetrated (Nielson and Don, 1974). Similarly, aphids feeding on non-host plants initially ingest phloem sap at normal rates, but then suddenly cease feeding, withdraw their stylets and leave the plant (Kloft, 1977). Such observations have led to the suggestion that the phloem sap of some resistant and non-host species is nutritionally unsuitable. However, grafting does not influence the growth and reproduction of *Aphis gossypii* on either resistant or susceptible muskmelon plants. The aphid still does ten times better on the 'susceptible' part of a graft, be it the scion or the stock, than on the 'resistant' part. Since neither resistance nor susceptibility is translocated across a graft union and the aphid finds no difficulty in locating the phloem of resistant plants, it is thought that other parts of the phloem such as the companion cells or cell walls produce a phytoalexin on contact with aphid stylets or saliva (Nielson and Don, 1974; Kennedy and Kishaba, 1977).

Secondary plant substances have been viewed as 'flags' by which host-specific aphids recognize their host plants (van Emden, 1978). For example, the cabbage aphid (*Brevicoryne brassicae*) will not settle and feed on plants lacking sinigrin (Figure 2.8) but can be reared on leaves of broad bean provided they are impregnated with sinigrin (Wensler, 1962). However, although it does not deter colonization (Figure 2.8), sinigrin has a negative effect on the growth of the polyphagous aphid *M. persicae* (van Emden, 1972) (Figure 2.9). High concentrations of the flavanoids catechin and epicatechin in the late buds and flowers of rose are also thought to deter the rose aphid (*Macrosiphum rosae*) from feeding on these structures (Miles, 1978). The

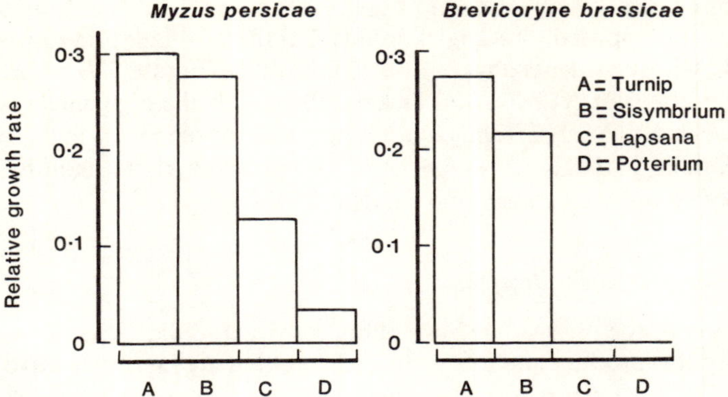

Figure 2.8 The relative growth rate of *Myzus persicae*, a generalist, and *Brevicoryne brassicae*, a specialist aphid, on plants that have (A + B) and lack (C + D) the secondary plant substance sinigrin. (After van Emden, 1972.)

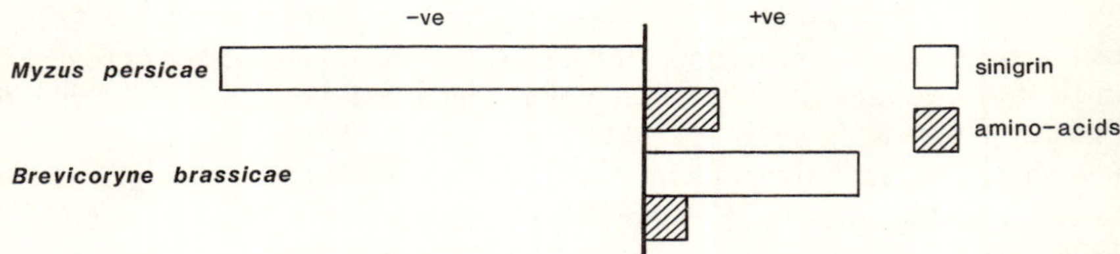

Figure 2.9 The relative effects of sinigrin and amino-acid concentration on the growth of a generalist aphid, *Myzus persicae*, and the specialist *Brevicoryne brassicae*. (After van Emden, 1972.)

broom aphid (*Aphis cytisorum*) prefers to feed on the flowering stems of broom, which contain more quinolizidine alkaloids than other parts of the plant (Figure 2.10), but tends to infest only the plants that have a below average alkaloid content (Wink *et al.*, 1982). Similarly, the hydrogen cyanide present in cyanogenic plants of white clover (*Trifolium repens*) is thought to explain why fewer *Aphis craccivora* are on these plants than on acyanogenic phenotypes of clover (Dritschilo *et al.*, 1979); high levels of phenol glucosides could account for the low suitability of some poplar leaves for galling by *Pemphigus betae* (Zucker, 1982); the high alkaloid content of 'bitter' lupins probably makes them resistant to aphids (Figure 2.11) (Węgorek and Krzymańska, 1971; Wink *et al.*, 1982); and the high levels of hydroxamic acids in the early developmental stages of certain grasses account for the poor population development on them of *Metopolophium dirhodum* (Figure 2.12), *Rhopalosiphum maidis* and *Schizaphis graminum* (Long *et al.*, 1977; Argandona *et al.*, 1980, 1983).

Although the presence of secondary plant substances in tissues other than the phloem is well documented, there is little, but growing, evidence to indicate that these substances or their precursors are to be found in phloem

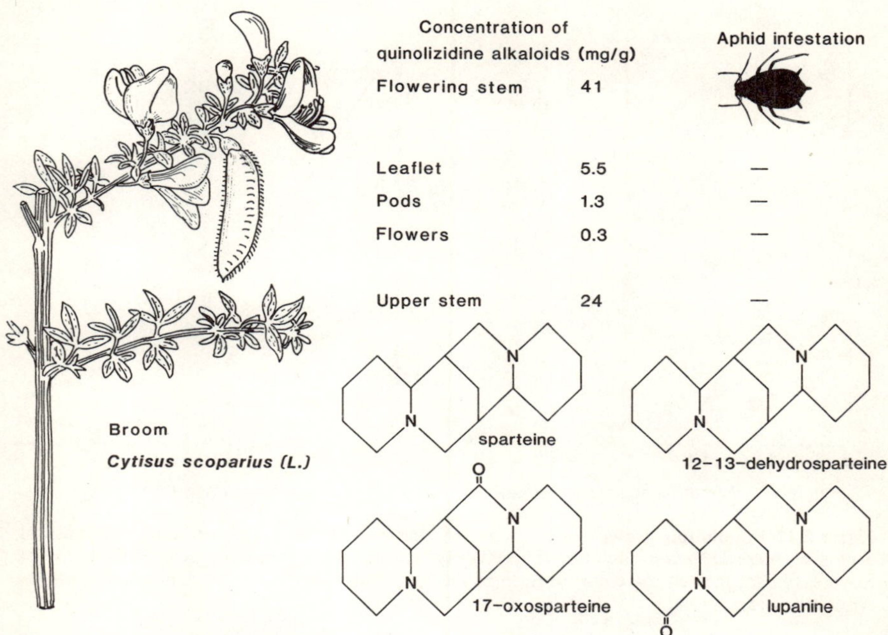

Figure 2.10 The total concentration and distribution of the quinolizidine alkaloids, sparteine, 12-13-dehydrosparteine, 17-oxosparteine and lupanine in broom in relation to aphid infestation. (After Wink *et al.*, 1982.)

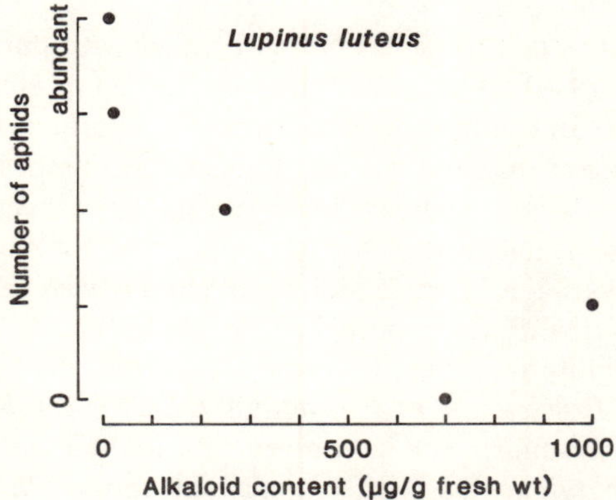

Figure 2.11 Abundance of aphids on lupin in relation to their alkaloid content. (After Wink *et al.*, 1982.)

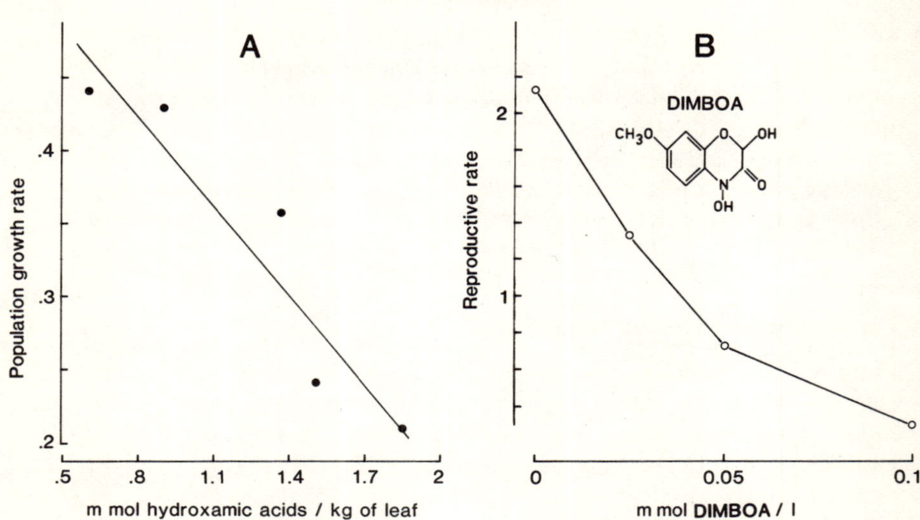

Figure 2.12 Population growth rate on cereals and reproductive rate on synthetic diets of *Metopolophium dirhodum* in relation to (*A*) the concentration of hydroxamic acids in the leaves of its cereal hosts and (*B*) the concentration of the hydroxamic acid DIMBOA in the diet. (After Argandona *et al.*, 1980.)

sap (Eschrich, 1970; Wink *et al.*, 1982). Certain defensive chemicals like the non-protein amino acid canavanine and phenols have been recorded from phloem sap (Ziegler, 1975; Dixon, 1975*a*). The oleander aphid (*Aphis nerii*) even sequesters and uses in its own defence the cardenolides that it imbibes from its host plants *Nerium oleander* and *Asclepias curassavica* (Rothschild *et al.*, 1970). Similarly, the broom aphid sequesters the four alkaloids produced by its host plant (Wink *et al.*, 1982).

M. persicae is probably the most polyphagous of all aphids, yet it has not been recorded from even 1% of all vascular plants (Eastop, 1973) and particular genotypes of this aphid are associated with particular species of plants (Takada, 1979). This aphid has a marked preference for highly nutritious plants (van Emden and Bashford, 1971) perhaps because the small quantities of toxicant ingested when aphids feed on tissues rich in nutrients may be quickly detoxified, whereas the larger quantities ingested when feeding on poor-quality hosts may not, as has been suggested for *Eriosoma* feeding on various resistant varieties of apple (Sen Gupta and Miles, 1975). Therefore, in selecting for high-quality food, polyphagous aphids like *M. persicae* possibly overcome the debilitating effect of some secondary plant metabolites. The specialist aphids *Brevicoryne brassicae* and *Lipaphis erysimi* feed on cruciferous plants rich in sinigrin and have an enzyme in their tissue (glucosinolase) capable of detoxifying this highly toxic substance (MacGibbon, 1975; MacGibbon and Beuzenberg, 1978). However, although it remains to be shown that glucosinolase protects these aphids from the toxic effects of sinigrin, it is possible that detoxifying enzymes have enabled many host specific aphids to circumvent the chemical defences of their host plants. Therefore it is likely that by feeding on phloem, aphids do not escape ingesting or coming into contact with a plant's defensive chemicals, and so an aphid's ability to detoxify them, or avoid inducing their production, has been important in restricting the distribution of aphids both between different genotypes of their host plants and more importantly between different species of plants. In most cases there is only one or a few species of plants an aphid can stomach.

Distribution between and on host plants

Phloem sap is rich in sugars but relatively poor in amino acids, which are essential for growth, and so aphids have to ingest very large amounts of sap in order to acquire sufficient protein. For example, an adult sycamore aphid requires 2.1 μl of sap per day, the contents of 5 200 sieve elements, and even a first instar requires 0.8 μl per day or the contents of 2 000 sieve tubes (Dixon and Logan, 1973). Most of the sap is excreted as droplets of honeydew which provides a rich source of sugar for many insects but especially ants.

Most insects that live on nutritionally-unbalanced diets possess symbiotic micro-organisms (Trager, 1970). In aphids, the symbionts are confined to special groups of cells called mycetomes, which are particularly well developed in nymphs. Tóth (1940) suggested that aphids utilize their symbionts to supplement their poor-quality diets, but the willow aphid (*Tuberolachnus salignus*) can obtain all the nitrogen it needs from the food it ingests (Mittler, 1958*b*). However, the slow growth and partial or complete sterility of experimentally produced aposymbiotic (i.e. symbiont-free) individuals indicates that symbionts can be important to aphids (Houk and Griffiths, 1980) for the supply of various substances that are absent or in short supply in phloem sap. For example the symbionts of *Neomyzus circumflexus* incorporate $^{35}SO_4$ into sulphur amino acids and ^{14}C-acetate into sterols (Ehrhardt, 1968); and those of *Acyrthosiphon pisum* may incorporate ^{14}C-acetate ^{14}C-mevalonate into a range of lipids, including exceptionally high levels into the sterol cholesterol (Houk *et al.*, 1976; Campbell and Nes, 1983). Aposymbiotic nymphs grow better on synthetic diets than on plants, which suggests that synthetic diets have fewer nutritional deficiencies and imbalances than phloem sap, even though sap provides a better food for the growth of aphids harbouring symbionts. The association aphids have evolved with their symbionts has possibly enabled them to exploit phloem sap (Mittler, 1971).

Food quality

The growth and reproduction of aphids is dependent upon the state of growth, or level of soluble nitrogen, in their host plants (Kennedy *et al.*, 1950; Mittler, 1958*a*; Dixon, 1970*a*). There is more nitrogen in the phloem sap of plants whose leaves are growing or senescent, because nutrients are then being actively translocated into or out of the leaves. The sap is not as nutritious when the leaves are mature. Thus there are marked seasonal changes in the quality of food available to aphids feeding on a particular plant and at a particular time the quality of food available from different plants varies. This is an important factor determining the distribution of aphids between plants (Kennedy *et al.*, 1950).

Aphids do not merely imbibe phloem sap. They simultaneously secrete substances into plants that rapidly affect the plant, in many cases to the aphids' advantage. Both the black bean aphid (*A. fabae*) and the cabbage aphid (*Brevicoryne brassicae*) can change the metabolism of their host plants, and although there are no outward signs, there are local changes in the plant around the feeding area that result in the development of larger and more fecund aphids (Way and Cammell, 1970; Dixon and Wratten, 1971). A few aphids, like *Myzus ligustri* and *Periphyllus acericola*, induce a localized yellowing of the tissues of mature leaves, which also makes the tissues attractive to other species of aphids, which feed on the same hosts but cannot induce senescence.

Some aphids are able to induce plant tissues to grow around and enclose them in complex and often colourful galls. As well as protecting aphids from the weather and general predators, galls offer a richer food supply than ungalled leaves and, very significantly, the production of galls also extends the usual period for which high-quality food is available (Forrest, 1971). The shapes of the very different galls induced by different species of *Pemphigus* on poplar are determined by the probing patterns of the aphids that initiate the galls (Dunn, 1960). Even the position of the ostiole, through which the aphids subsequently leave, is determined by the aphid making 10 times as many probes in the area immediately around the future ostiole than in the area of the ostiole itself (Schwarzbach, 1962). The ostiole opens when the gall begins to dry out (de Geer, 1773) and will close when dampened (Frank, 1896), as a consequence of changes in the water content of the intercellular pectic material (Rohfritsch, 1966).

Miles (1968*a*, *b*) proposed that the form of the gall can be attributed to a single chemical substance, indole acetic acid, present in the saliva, and the behaviour of the insect and the nature of the plant tissue attacked are also important. Certainly the behaviour of the gall-forming aphid is important in aphids such as *Pemphigus*. *Dysaphis devecta* or *D. plataginea*, if they are confined to the stem of an apple seedling, can induce leaves several centimetres away to develop leaf galls characteristic of each species, but it is unlikely that the marked differences in form and colour of these leaf galls on the same host are attributable to a single substance. More probably the saliva differs in each species and therefore in the way it affects leaf development (Forrest and Dixon, 1975) and improves the quality of the phloem sap for the aphids.

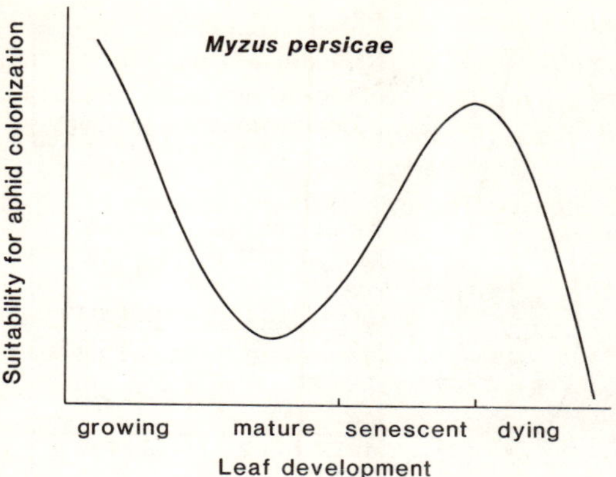

Figure 2.13 Relative suitability for colonization by *Myzus persicae* of leaves of successive stages of development. (After Kennedy *et al.*, 1950.)

Feeding site

Not all the parts of a plant are equally suitable sources of food for aphids. The tendency to move more frequently when feeding on a poor than on a good food source, along with their geotactic and photactic responses, results in aphids colonizing parts of plants on which they can achieve the highest growth and developmental rates (Kennedy *et al.*, 1950; Ibbotson and Kennedy, 1950) (Figure 2.13).

That the best feeding sites are often in short supply has been elegantly shown by Whitham (1979) who studied leaf selection in galling aphids. *Pempigus betae* produces the largest and most fecund galls on the basal part of the largest leaves of cottonwood (*Populus angustifolia*). There are fewer large leaves than small leaves, and when two stem mothers (p. 46) attempt to settle on the same leaf, a kicking and shoving contest follows that can last for two days. The larger aphid usually wins, displacing the loser to a more distal position along the midrib. However, the loser's chances of initiating a gall in this position are less and, even if successful, it will have fewer descendents than the aphid in the prime basal position. The potential genetic fitness of a third aphid attempting to colonize a leaf is even lower (Figure 2.14), and so by means of territorial behaviour, large aphids can deny other individuals access to the best leaves (Whitham, 1979).

In *Epipemphigus niisimae* there is also evidence of a shortage of highly suitable leaves for galling. When there is a large number of galls on a twig, there are more likely to be one or more dead stem mothers (intruders) in some of the galls than when there are few galls per twig (Figure 2.15). Like *P. betae, E. niisimae* also fights, this time for possession of a gall, and a large individual

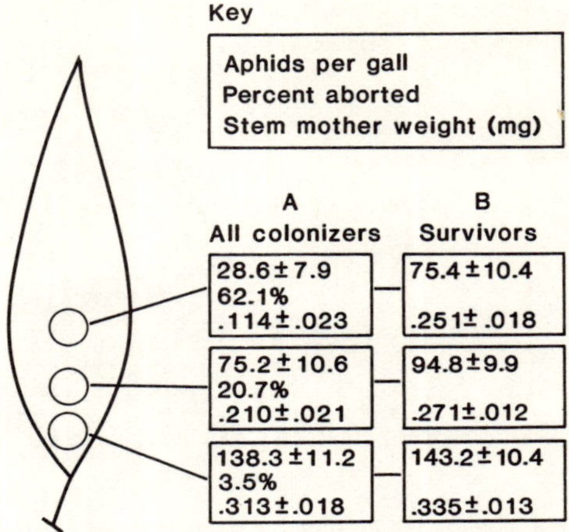

Figure 2.14 Effect of gall position on relative fitness when three *Pemphigus betae* stem mothers colonize the same leaf. (After Whitham, 1978.)

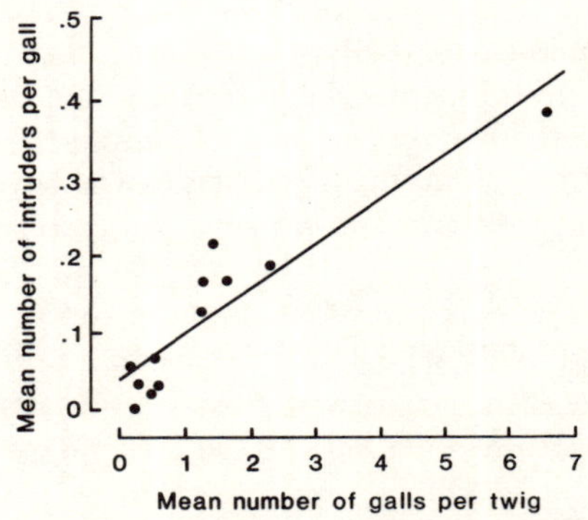

Figure 2.15 Mean number of additional (intruder) stem mothers per gall of *Epipemphigus niisimae* in relation to the mean number of galls per twig. (After Aoki and Makino, 1982.)

is capable of using its stylets to stab and kill a smaller intruder (Aoki and Makino, 1982).

For *P. betae* the size of a leaf reflects its quality. However, each additional aphid founding a gall on a leaf, apart from having a lower genetic fitness, also reduces the fitness of the aphid(s) already present. As a consequence, the average genetic fitness of aphids galling a leaf declines with an increase in the number of competitors. A large leaf, 14.6 cm long, with three galls, will sustain the equivalent number of aphids per gall as a medium leaf of 12.3 cm, with two galls, or a small 10.2 cm leaf with one gall (Figure 2.16). The observed relationship between number of galls per leaf and leaf size follows this closely,

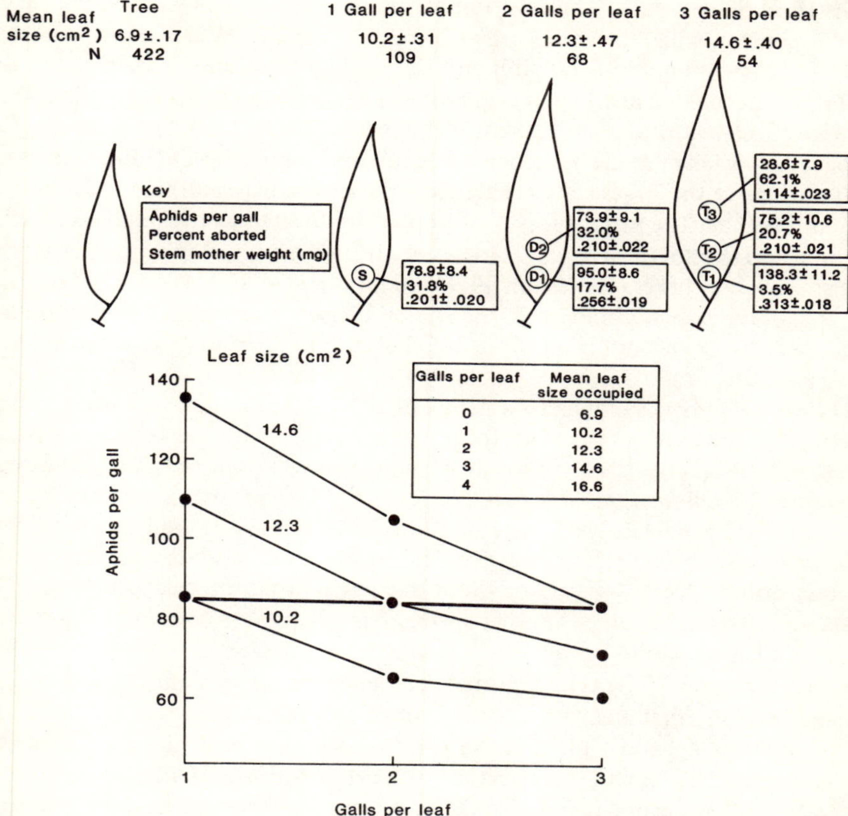

Galls per leaf	Mean leaf size occupied
0	6.9
1	10.2
2	12.3
3	14.6
4	16.6

Figure 2.16 Distribution of galls and estimated fitness of *Pemphigus betae* in relation to leaf size and number of galls per leaf. S = single gall per leaf; D = double galled; T = triple galled; N = sample size. See text for details. (After Whitham, 1980.)

showing that the average fitness of aphids on a leaf is independent of leaf size (Whitham, 1980). This appears to agree with the ideal free distribution hypothesis put forward by Fretwell and Lucas (1969) which postulates that as population density increases, the habitat patches should be occupied in a sequence that reflects the ranking of their value as feeding sites. That is, an aphid should gall a small leaf, or form the second gall on a medium-sized leaf, rather than the fourth gall on a large leaf. It is individual fitness that is important, however, not average fitness. If efficient at maximizing their fitness, then it is puzzling why aphids that initiate the third gall on a large leaf do not colonize uninfested small leaves, which are just as abundant, and on which they could give rise to three times as many descendants.

Dunn's (1960) detailed observations on the rapid gall initiation in other species of *Pemphigus* should also be borne in mind when interpreting the distribution of galls. Aphids initiate galls in the early stages of leaf enlargement, when the leaves are growing very rapidly. When selecting a leaf at this stage an aphid is possibly not responding to its size or even nutritive status, but the rate at which it is growing. The faster the tissue is growing, the more effective an aphid is at galling it (Forrest and Dixon, 1975). Generally, as leaves grow faster at the base than at the tip, the best position to initiate galls is at the base of the larger leaves. Then, too, the leaves do not all emerge from the bud together. The largest leaves of poplar are located at the middle of each shoot and are not the first leaves to emerge from the buds. Poplar aphids hatch just *after* bud burst (Whitham, 1978). As in other species, the timing of egg hatch of the poplar aphid (p. 72) has possibly been determined by selection to coincide with the unfurling and initial very rapid growth of what are to become the largest leaves.

Eriosoma yangi belongs to a genus of aphids that commonly form galls. However, this species does not form a gall, but feeds on the young leaves of elm until it reaches the third instar when it then seeks out the gall of another species of *Eriosoma.* It is able to kill the owner of the gall and take it over for its own use. *E. yangi* has been found in the galls of as many as four other species of *Eriosoma* and therefore does not depend upon the availability of only one other aphid species (Akimoto, 1981). It appears that under certain circumstances to usurp a gall of another species has advantages over inducing one's own gall. This would be favoured by an abundance of galls, and intraspecific gall usurpation of the type reported for *Epipemphigus niisimae* (p. 19) could have been the first step in the evolution of gall parasitism.

Summarizing, most plants have evolved chemical and physical defences that operate at different stages of the host selection process. Chemical defences are possibly widespread and difficult to overcome and may in part account for the relative rarity of polyphagy in aphids. By confining their attacks to certain plants when they are a rich source of food, polyphagous aphids may be able to compensate for the cost of detoxifying the chemical defences. Monophagous aphids have possibly evolved detoxifying enzymes or other appropriate mechanisms, whilst some use the toxic chemical as a flag by which to recognize their host plant and some even sequester the chemical for their own defence. Phloem sap is a poor food because of its low nitrogen content. However, aphid symbionts possibly supplement the diet by supplying certain essential nutrients either lacking or in short supply in phloem sap. Certain aphids can also modify plant metabolism and development to their own advantage and all show a marked tendency to colonize only the nutritionally richest parts of their host plants.

3 Size in aphids

Small size has been a feature of aphids ever since their evolution in the Permian. Even so, there is a range of size within and between species. In general, bigger organisms have a better chance of reaching a reproductive condition and then tend to produce more gametes (Calow, 1978). Therefore it is of interest to consider what determines size in aphids.

Size within a species

Adults of several species show as much as a 10-fold range in size (Way and Banks, 1967; Murdie 1969*a*, *b*; Dixon and Dharma, 1980*a*). Small individuals develop when they are reared in crowded conditions, on mature plants or at high temperatures; large individuals result when nymphs are reared in isolation, on young or senescent plants or leaves, or at low temperatures. Well-fed animals generally grow larger than poorly-fed ones; therefore the effect of food quality on size is not surprising. An inverse relationship between size and temperature is widespread in insects.

Müller (1966*a*) postulated that size in aphids is a consequence of the balance between catabolism and anabolism. At high temperatures catabolism consumes most of the available energy, and little is left for growth (anabolism). Maximum size is achieved at the temperature at which the energy needs for catabolism are minimal. Similarly, an imbalance in the energy requirements of basic metabolism and growth has also been proposed to account for the smallness of certain Lepidoptera when reared at high temperatures (Marthavan and Pandian, 1975). Small size in *Myzus persicae* at high temperatures has been attributed to an adverse effect on the aphid's symbionts (Lawson, 1941) and resultant deterioration in the quality of the food available to the aphid.

By treating development and growth as separate processes, Chambers (1979) showed that although aphids reared at high temperatures are small, they nevertheless have a higher growth rate than those reared at low temperatures (Dixon *et al.*, 1982). Size is a consequence of the relative effect of food quality and temperature on the growth and developmental rates. Both an

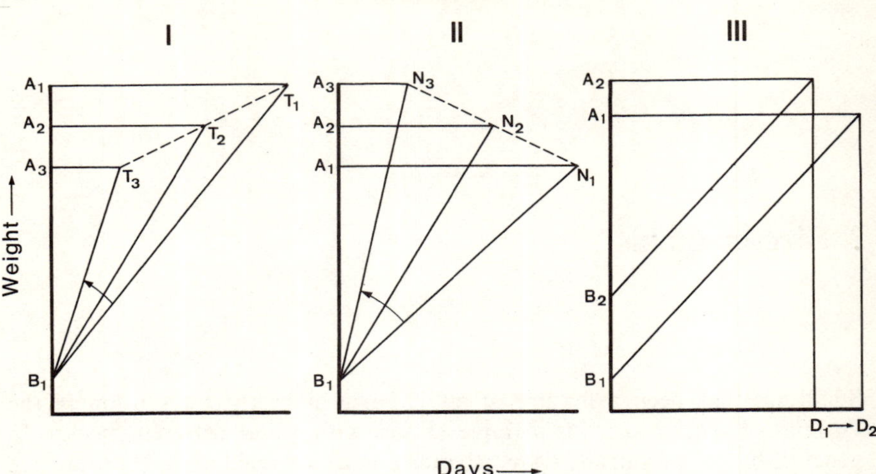

Figure 3.1 The relationships between adult weight (A_1, A_2, A_3) and developmental times for: aphids reared on food of the same quality at a range of temperature (T), where $T_1 < T_2 < T_3$, (I); at one temperature and a range of food qualities (N), where $N_1 < N_2 < N_3$, (II); and at one temperature and a particular food quality but starting with nymphs of different birth weights (B_1, B_2), where $B_1 < B_2$. and D_1 and D_2 are the developmental times (III). Weight is plotted on a logarithmic scale. The relative growth rate ($\overline{RGR}$) in each case is the slope of the relationship between weight and developmental time.

increase in food quality and temperature result in an increase in the growth rate measured as the increase in weight per unit weight per unit time ($\overline{RGR}$). However, increases in temperature disproportionately decrease the time it takes to reach maturity (Figure 3.1, I). Similarly the time to maturity is affected by an increase in food quality, but the effect is subproportional (Figure 3.1, II). As a consequence aphids are small when reared at high temperature and large when reared on high-quality food.

Time to maturity is also affected by birth weight so that the generally small nymphs born to small mothers take longer to mature than the large nymphs born to large mothers (Figure 3.1, III). However, the increase in developmental time does not compensate completely for the difference in weight at birth to produce a generation of adults uniform in size. Usually for a particular set of conditions a clone of initially large aphids will decrease in size and a clone of small aphids will increase in size over a number of generations, converging on a particular size (Figure 3.2). The final size is a consequence of the effect of temperature and food quality on the developmental and growth rates. A lower temperature or higher food quality increases the final size, and a higher temperature or lower food quality decreases the final weight.

That aphids small at birth are slow to reach maturity may simply reflect the early stage of development at which these aphids are born. Attempts to correct for this have proved unsuccessful (Dixon *et al.*, 1982). If we ignore the effect of

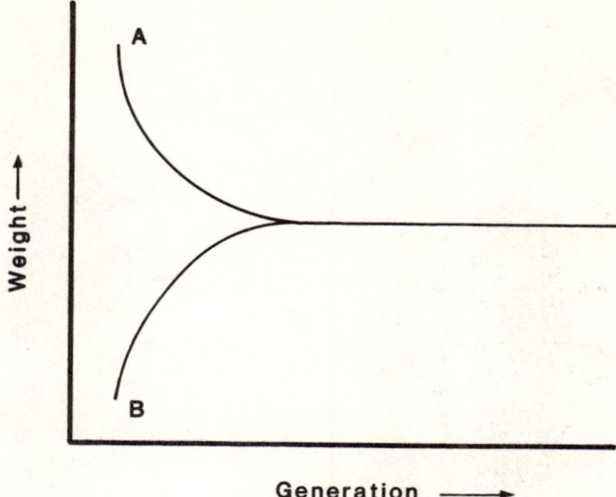

Figure 3.2 The trend in adult weight from generation to generation of initially large (*A*) and small (*B*) aphids reared at the same temperature and on food of the same quality.

birth weight on the developmental rate, it could be argued that both growth and development occur at a maximum rate. It is likely, however, that the magnitude of the growth and developmental rates is also determined by selection, as in other animals where it has been possible to select for changes in these life-history parameters (Doyle and Hunte, 1981).

On being the right size

Different species vary in length from 0.7 to 7 mm, and those that feed on the deeply located phloem elements in the trunks of trees are larger than those that feed on the more accessible phloem elements of leaves. To reach phloem elements deep within a trunk of a tree an aphid needs long stylets. They are not coiled up within the body, as in scale insects, but are characteristically enclosed within a proboscis. The stylets are exposed by telescoping the distal segments within the basal segments of the proboscis (Figure 2.3). *Stomaphis quercus* feeds on the trunks of mature oak trees and has a proboscis that is nearly twice the length of its body. The telescoping of the proboscis, necessary to expose the stylets, results in the invaginated proboscis running internally, the full length of the body (Figure 3.3). Thus the structure of the proboscis that evolved as long ago as the Jurassic (Shaposhnikov, 1977) makes it physically impossible for small aphids to feed on tissues very deep within a plant.

The range in size, with large species of aphids feeding on the trunks of trees and small species of aphid feeding on leaves, is well illustrated by the aphids feeding on oak (Figure 3.4). Similarly, the generations of *Periphyllus* that feed on the twigs of sycamore are larger and have a proportionately longer proboscis, than the generations of the same species that feed on the leaves of

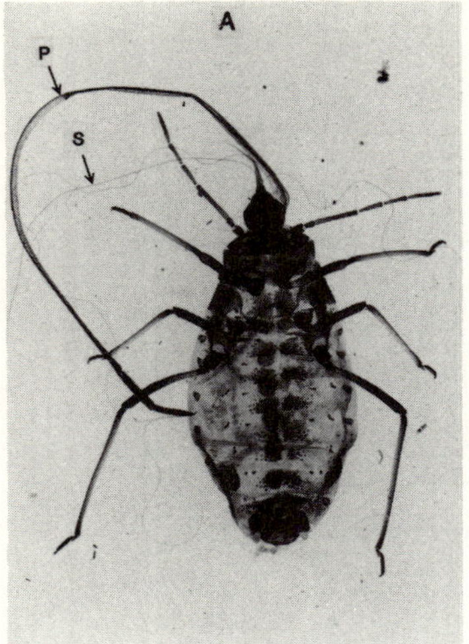

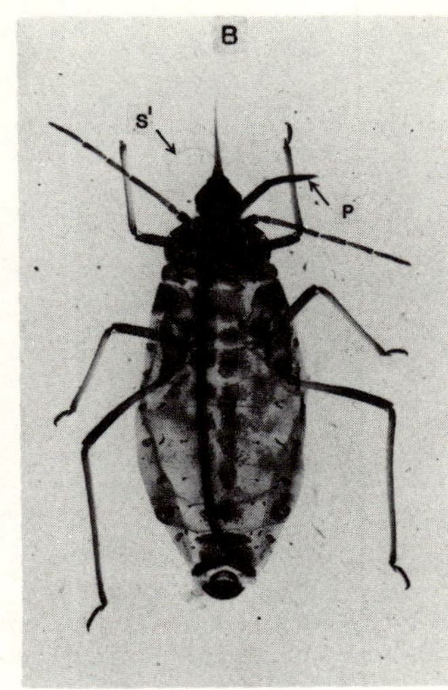

Figure 3.3 Photographs of slides of *Stomaphis quercus* showing its long proboscis and fine stylets (*A*); and the proboscis in the feeding position, invaginated on itself as far back as it can go within the abdominal cavity (*B*). (P = proboscis, S = stylet, S^1 = base of cut stylet).

sycamore (Figure 6.11). On birch leaves the smallest aphid, *Callipterinella minutissima*, feeds on the smallest veins; the intermediate-sized *Betulaphis quadrituberculata* feeds on medium-sized veins; and *Euceraphis punctipennis*, the largest aphid, feeds on the largest and deepest veins. Within a species, successive instars feed on larger veins of a leaf (Figure 3.5) (Dixon and Logan, 1973). Thus the size of an aphid is correlated with the depth of the phloem elements on which it feeds.

By feeding on minor, rather than major, leaf veins aphids have the advantage of a shorter pathway to the sieve tubes, less sclerenchyma to impede the passage of their stylets and also a food supply richer in both sugars and proteins (Dixon and Logan, 1973). The question therefore arises: why don't all leaf-dwelling aphids feed on the rich sap of the minor veins? The answer possibly lies in the fact that large aphids remove more sap per unit time than small aphids. Therefore large aphids are more likely to induce velocities of sap flow in the minor veins that exceed the rate normally tolerated by the plant, and beyond which sealing of the phloem element occurs, as in a wounding response. This also applies to the large species of aphids that feed on the large phloem elements deep within the branches and trunks of trees, and the plant's wounding response may have prevented these aphids from becoming even larger.

Another factor that may restrict the development of large, leaf-feeding aphids is that their size and weight would make them more prone to being dislodged or disturbed during their feeding. However, although the shape of aphids living protected within rolled leaves and leaf galls tends to be more

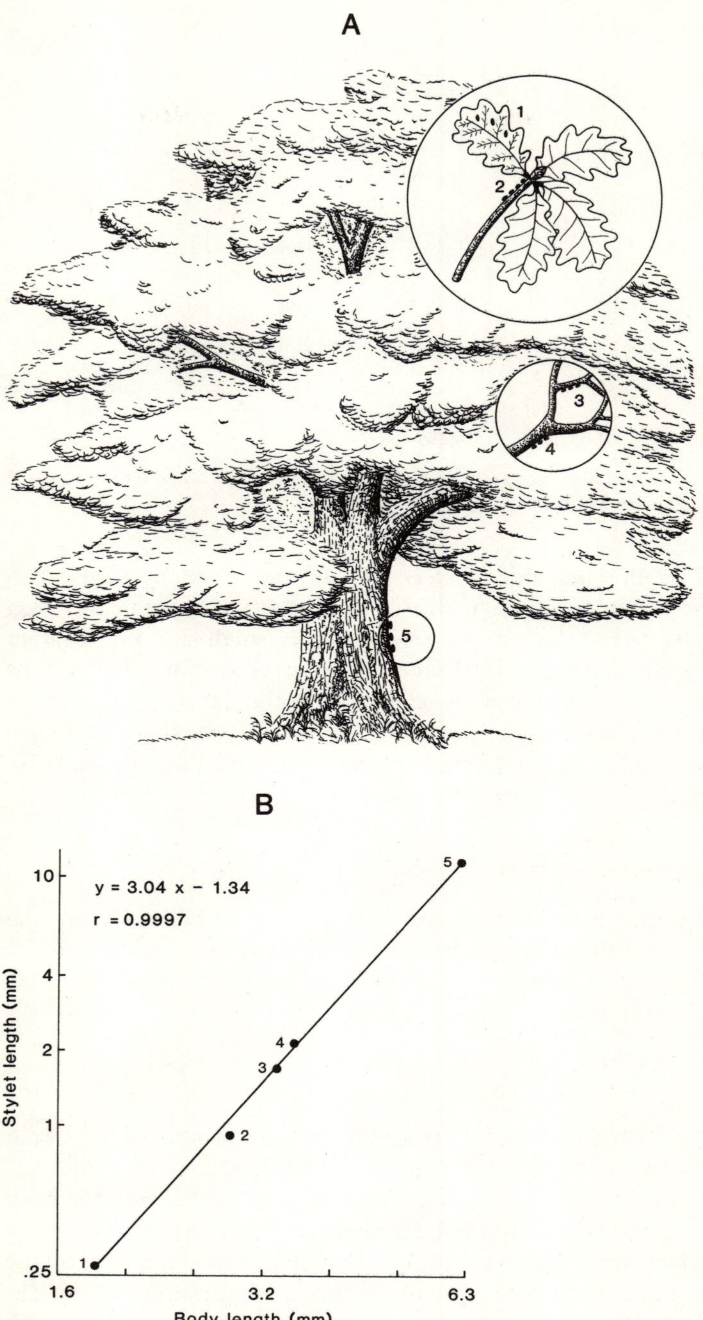

Figure 3.4 The feeding positions of six species of aphids living on oak (*A*), and the relationship between the length of the stylets and their body plotted on a logarithmic scale (*B*). 1, *Tuberculoides annulatus*; 2, *Thelaxes dryophila*; 3, *Lachnus roboris*; 4, *Lachnus iliciphilus*; 5, *Stomaphis quercus*.

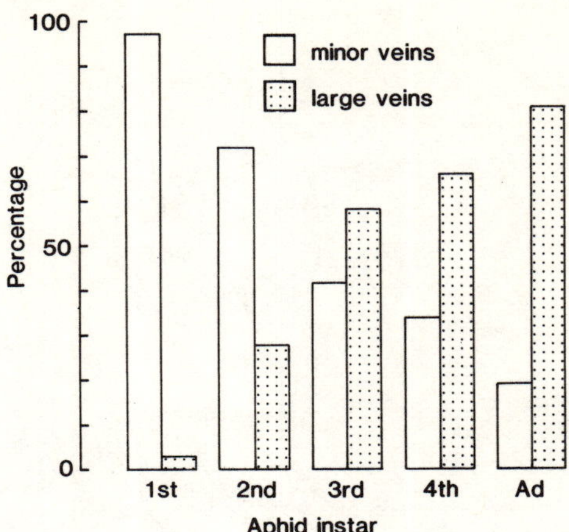

Figure 3.5 The percentage of first, second, third, fourth and adult instars of the sycamore aphid feeding on minor and large leaf veins.

spindle-shaped or globular, nevertheless they are not generally larger than other leaf-feeding aphids. Although Heie (1967) suggested dispersal by riding the winds to have been a factor in the evolution of small size, much larger insects such as locusts (Rainey, 1963) and butterflies (Johnson, 1969) use the winds in this way. Thus both the phylogenetic constraint imposed by the mode of feeding, and the physiological constraint imposed by feeding on phloem elements have to a large extent determined the small size of aphids, and also the size differences between particular species of aphids.

Size and proportions

A small aphid is not a small duplicate of a large one. Not only are the proportions of a first instar aphid different from those of an adult, but the proportions of a small species are different from those of a large one. This affects feeding, locomotion and reproduction.

Feeding

As the proboscis of large species is longer than the legs, projecting back even beyond the end of the abdomen, large species cannot feed as small aphids do, by swinging the proboscis forward into a vertical position beneath the head and then inserting the stylets. In the large species aphids have to back onto their stylets, which are then exposed by telescoping the segments of the proboscis. This is even more striking in the offspring of large aphids, which at birth have an even longer proboscis in proportion to the body (Figure 3.6).

Small species can withdraw their stylets in seconds, whereas *Stomaphis*

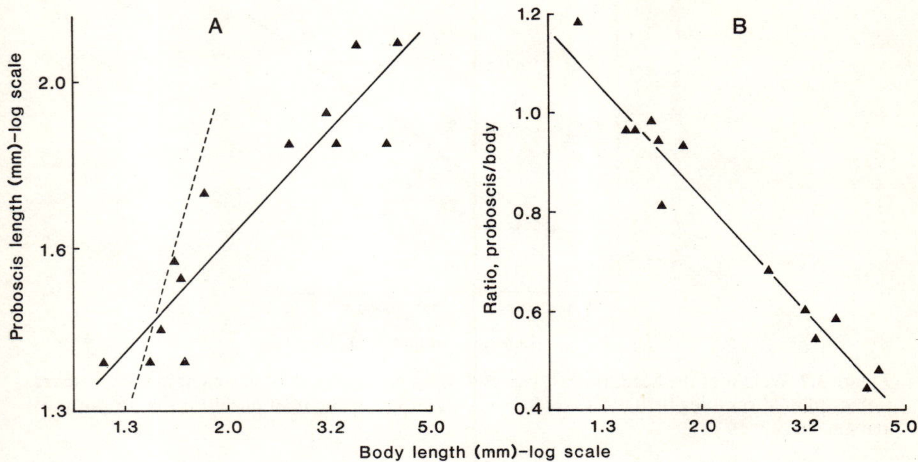

Figure 3.6 Proboscis length (*A*), and ratio of proboscis length to body length (*B*) in relation to body length during growth from first instar to adult in the large willow aphid (*Tuberolachnus salignus*). (--- if the slope were equal to 1).

quercus can take as long as 50 minutes during which time it is vulnerable to attack. In this context, it is interesting to note that large aphids tend to be ant-attended, which affords them some protection from general predators.

Locomotion

In alate aphids the body can be thought of as comprising a vegetative component (the abdomen, which contains most of the alimentary canal and the gonads), and a combined locomotor and sensory component (made up of the head, thorax, legs and wings). In small sycamore aphids the latter component is a relatively larger proportion of the total weight (Figure 3.7). In addition wing loading is lower and wing beat frequency higher than in large individuals (Dixon, 1974, 1975*b*; Mercer, 1979) and similar trends occur in other species of aphids. As aphids have very little control of their flight direction, the ability to take off and fly upwards and the time of take-off (p. 95) are more important than their horizontal flight speed.

The vertical flight speed in cm s^{-1} (y) in sycamore aphids is correlated with both wing length in mm (x_1) and temperature in °C (x_2):

$$y = 17.1x_1 + 3.0x_2 - 98.2$$

Thus when resources are limited there are advantages in putting more of them into developing a larger thorax and longer wings, and less into reproduction,

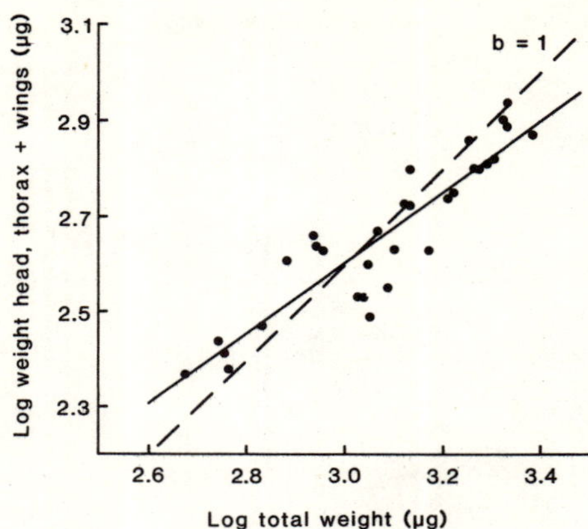

Figure 3.7 Weight of the head, thorax, legs and wings in relation to total weight in the sycamore aphid, plotted on a logarithmic scale. ($b = 1$ is the slope of the relationship if the proportions remain the same).

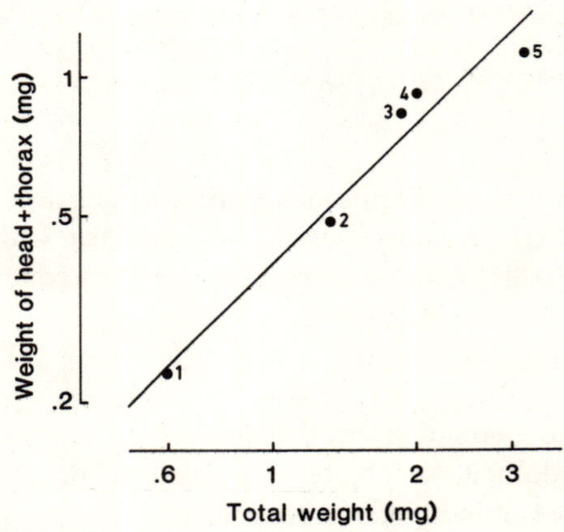

Figure 3.8 Weight of the head, thorax, legs and wings in relation to total weight for five species of aphid. 1, *Rhopalosiphum padi*; 2, *Drepanosiphum platanoidis*; 3, *Callaphis juglandis*; 4, *Megoura viciae*; 5, *Tuberolachnus salignus.*

so as to remain flightworthy. Very short-winged individuals cannot easily fly upwards, especially at low temperatures (Mercer, 1979).

The proportion of the body that on average makes up the locomotory apparatus is only known for a few species (Figure 3.8). However, for these aphid species it is approximately 40%. Therefore, on the basis of a few species, large ones do not assign proportionately more or less of their resources to locomotion than small ones.

Small species have shorter wings and a higher wing-beat frequency than larger ones. For individuals of the black bean, lime and sycamore aphids, that have wings ranging in length from 2 to 6 mm, the relationship between wing-beat frequency in cycles s^{-1} (y) and wing length in mm (x) is:

$$y = 153.8 - 16.7x$$

The aphids with the shortest wing lengths have the highest wing-beat frequency (Mercer, 1979). Artificially shortening the wings of sycamore aphids results in a dramatic increase in their wing-beat frequency. This is characteristic of insects with asynchronous flight muscles, in which the frequency of contractions of the flight muscles is much greater than, and not closely related to, the frequency of motor nerve impulses.

When a host plant deteriorates there are advantages to an aphid in investing more resources in locomotion and dispersing to other host plants. This is seen in both the switch to alate production (p. 38) and the trade-off between the investment in locomotion and reproduction with small individuals of a species investing more than large individuals in locomotion in order to achieve take-off. However, small species of aphids do not appear to invest more in locomotion than large species. In general, the smaller the aphid species the faster it beats its wings and the slower its flight (Johnson, 1969; Mercer, 1979).

Reproduction

The largest individuals in many species produce the most, and the largest, offspring (Figure 3.9). In the sycamore aphid this reflects the relative size of the abdomen (p. 29), which represents the proportion of the resources invested in reproduction during nymphal life. Large offspring are fitter than small offspring because they are more likely to survive harsh conditions (Figure 3.10), and avoid capture by predators (Figure 3.11), or parasitization by hymenopterous parasites (Mackauer, 1973). Thus, in general within a species, large individuals have a greater potential than small individuals.

Although large species produce large offspring, each offspring is a smaller

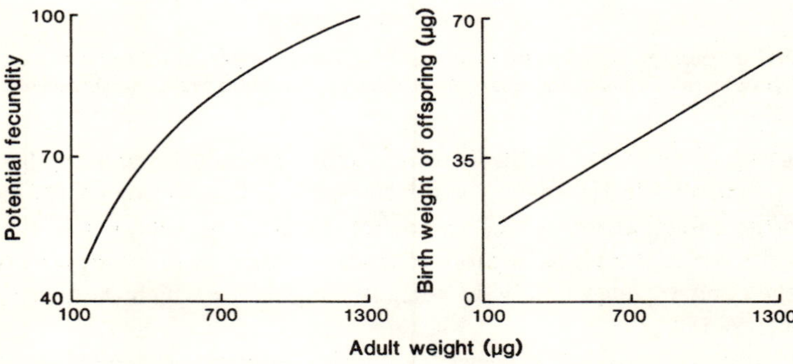

Figure 3.9 Potential fecundity and birth weight of offspring in relation to adult weight in *Aphis fabae*.

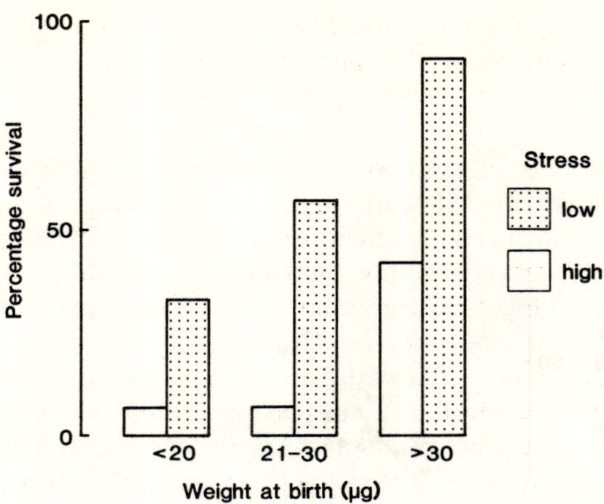

Figure 3.10 Percentage survival in relation to birth weight of *Aphis fabae* when reared on poor- and good-quality (unstippled, stippled) host plants, on which 88% and 44% died, respectively.

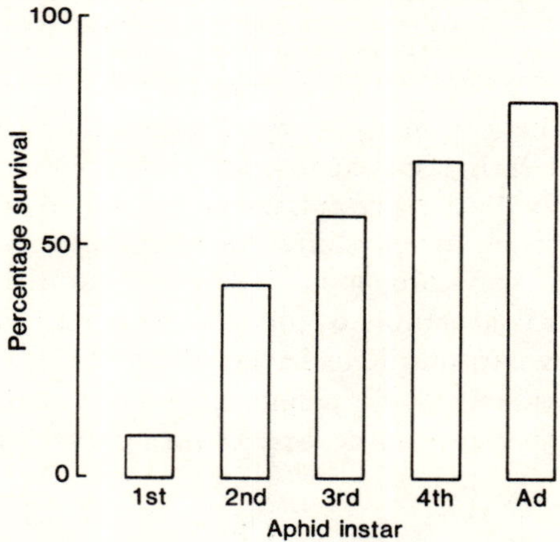

Figure 3.11 Percentage of the various instars of the nettle aphid, *Microlophium evansi*, that survive encounters with fourth instar larvae of the ten-spot ladybird beetle, *Adalia decempunctata*.

proportion (2%) of the mother's weight than it would be in a small species (10%) (Figure 3.12). However, the relative size of the abdomen is the same in both large and small species of aphids (cf. Figure 3.8). One would therefore expect the larger species to be more fecund, and there is some support for this from the twelve species of a wide range of weights for which there are results (Dixon, 1985).

The intrinsic rate of increase (r_m) of a species is dependent on its developmental rate and age-specific fecundity and survival schedules, and generally the rate of development is more important than fecundity and the

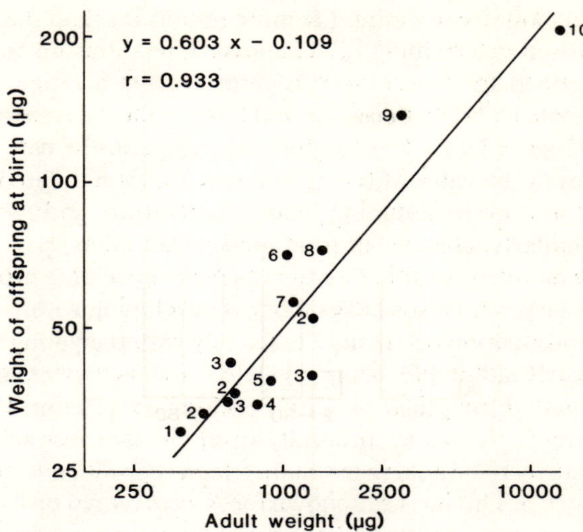

Figure 3.12 Offspring weight at birth in relation to adult weight for ten species of aphid, plotted on a logarithmic scale. 1, *Elatobium abietinum*; 2, *Rhopalosiphum padi* (fundatrices, emigrants, apterous exules); 3, *Aphis fabae* (fundatrices, apterous exules, alate exules); 4, *Drepanosiphum acerinum*; 5, *Periphyllus testudinaceus*; 6, *Sitobion avenae*; 7, *Metopolophium dirhodum*; 8, *Drepanosiphum platanoidis*; 9, *Megoura viciae*; 10, *Tuberolachnus salignus.*

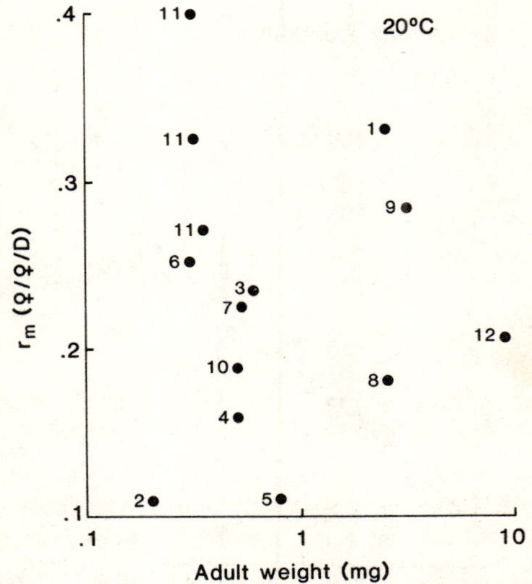

Figure 3.13 Intrisic rate of natural increase (r_m) at 20° C in relation to body weight for twelve species of aphid.1, *Acrythosiphon pisum*; 2, *Aphis citricola*; 3, *Aphis fabae*; 4, *Drepanosiphum acerinum*; 5, *D. platanoidis*; 6, *Elatobium abietinum*; 7, *Eucallipterus tiliae*; 8, *Macrosiphum albifrons*; 9, *Megoura viciae*; 10, *Myzus persicae*; 11, *Rhopalosiphum padi* on *Lolium perenne, Poa pratense* and *Dactylis glomerata*; 12, *Tuberolachnus salignus.* After Burns, 1971; Barlow and Dixon, 1980; Bintcliffe Wratten 1982; Campbell and Mackauer 1977; Dixon and Wratten, 1971; Fisher pers. com.; Frazer and Gill, 1981; Leather pers. com.; Llewellyn *et al.*, 1974; Wellings, 1981).

rate of reproduction in early adult life more important than the total number of nymphs born in determining r_m. The inverse relationship between r_m and body size described for heterothermic animals from a range of taxonomic groups by Fenchel (1974) does not hold for aphids, even at a constant temperature (Figure 3.13). Fluctuations in temperature can result in far greater changes in the value of the r_m of a species, than differences in weight will bring about, even assuming that temperature and weight can be dissociated. Similarly, changes in food quality can affect r_m.

On reaching maturity an aphid's embryos make up a large proportion of its weight, and its larger embryos have embryos developing within them. That is growth and reproduction occur simultaneously with the young that are to be produced in early adult life being ovulated and achieving most of their embryonic growth during their parents' development (Dixon, 1985*a*). Therefore irrespective of its size as an adult, an aphid that has achieved a high growth rate is likely to be able to maintain a high reproductive rate and a high rate of increase (r_m). This has been shown for *R. padi* reared on a range of hosts of different nutritional quality, and at several different temperatures (Leather and Dixon, 1984). It also holds for the six species for which there are measurements of the mean relative growth rate and r_m (Figure 3.14).

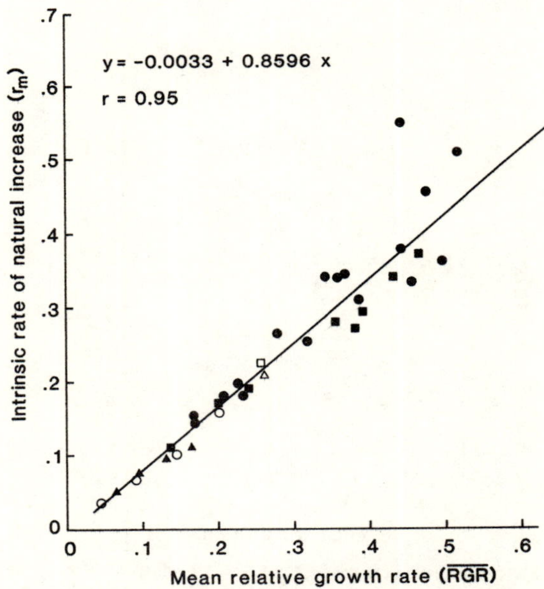

Figure 3.14 Intrinsic rate of natural increase (r_m) in relation to relative growth rate for six species of aphids reared under a wide range of conditions. ○, *Drepanosiphum acerinum*; ▲, *D. platanoidis*; □, *Eucallipterus tiliae*; ●, *Rhopalosiphum padi*; ■, *Sitobion avenae*; △, *Tuberolachnus salignus*. After Barlow and Dixon, 1980; Leather and Dixon, 1984; Llewellyn *et al.*, 1974; Watson, 1983; Wellings, 1981).

Summarizing, individual size is a consequence of the effects of nutrition and temperature on the developmental and growth rates, the values of which, like the proportion of the resources assigned to locomotion and reproduction, are shaped by selection. However, this is subject to two important constraints: the mechanism by which aphids expose their stylets to insert them into the plants, and the physiology of the phloem elements on which the aphids feed. These constraints account for the tendency of all aphids to be small, although those that feed on the woody parts of plants are larger than those that feed on leaves, because of the depth of the phloem elements. In favourable conditions aphids generally achieve high rates of growth and increase in numbers, irrespective of their weight, a feature characteristic of opportunistic species.

4 Polymorphism

An aphid life cycle is made up of a sequence of morphs that differ in both their behaviour and structure. In a few species there can be as many as eight morphs that differ at least in their external morphology. Polymorphism is characteristic of aphids and reaches its greatest development in host alternating species (p. 68). The ancestors of aphids were winged, as is still largely the case for many species in the subfamily Drepanosiphinae, in which only the sexual females lack wings. However, in many aphids the parthenogenetic females show alary dimorphism. That is to say they can be either winged (alate) or unwinged (apterous). Most individuals reproduce either asexually or sexually. The striking exception is the sterile soldier caste that has recently been recorded in a few species of the subfamily Pemphiginae (Aoki, 1975, 1977).

During periods of stress, such as mid-summer and winter, a few species can produce either aestivating or hibernating morphs. In autumn most aphids produce sexual forms: wingless females and usually winged males, although apterous males are known and a few species have both apterous and alate males. After mating, sexual females lay eggs and this enables many species to survive the winter and then resume development and hatch the following spring.

Individuals of each morph are involved to a varying degree in defence, dispersal, reproduction and survival. However, specialization in one or other of these functions imposes constraints in terms of resource allocation, physiology and structure for carrying out the other functions. At certain times particular functions are more important than others for survival, and this has resulted in the evolution of a division of labour within a clone that is reflected in its polymorphism.

Let us consider the adaptive significance of particular morphs and how their development is controlled. This can be viewed at the ecological and at the physiological level. A knowledge of both levels is necessary for a complete understanding of how aphids can rapidly adjust to, and to a very large extent anticipate, environmental change. It should be noted that the words anticipation, decision, investment, optimum, strategy, and tactics are not intended to imply anything about conscious thought. They are a short-hand way of saying that aphids showing particular responses and structures have been favoured by selection.

Defence

Individual aphids are capable of defending themselves against a wide range of natural enemies, with varying degrees of success. Depending on the relative size of the attacking parasite or predator, an aphid will kick, attempt to walk away, drop off the plant or smear the wax it can exude from its siphunculi over an attacker (Dixon, 1958, Figure 2). Waxing the attacker is often accompanied by the release of an alarm pheromone that alerts nearby aphids. There are several alarm pheromones. The most widespread interspecifically is trans-β-farnesene (Pickett and Griffiths, 1980). Aphids respond to these volatile substances by either swivelling their bodies, or vacating the area by walking, flying or jumping. However, in some aphids the alarm pheromone only serves to heighten the aphid's response to mechanical and visual stimuli associated with the presence of a parasite or predator (Dixon and Stewart, 1975; Nault and Montgomery, 1977). As adjacent aphids are likely to be of the same clone this altruistic act increases the probability of the clone surviving.

These defensive responses and the associated morphology may not be equally well developed in all morphs, especially in species that show a high degree of polymorphism. For example, only the first instar fundatrices of certain gall-forming aphids have the heavy sclerotization that affords them some protection against attack by other individuals of the species (Aoki and Makino, 1982).

Aoki's (1975, 1977) exciting discovery of a 'soldier' morph in *Colophina* spp. and *Pseudoregma* spp., is a good example of the division of labour within a clone, with some individuals specialized for defence. Soldiers can make up 13% of a colony, are very short-lived, do not feed or reproduce, and defend the colony against insect enemies (Figure 4.1). As a colony is likely to consist mainly of the descendents of a founder aphid (i.e., it is a clone), a soldier, in using its well developed prehensile forelimbs and frontal spike, or even its proboscis, to seize and kill potential insect enemies, reduces the probability of the clone's becoming extinct.

Approximately 50% of the 100–200 000 aphid inhabitants of the large galls produced by *Astegopteryx styracicola* on snowball trees are 'biters'. They emerge from the many exits in a gall and can be found on the adjacent twigs where they will attack predatory insect larvae. If a gall is handled the biters stream out and can inflict painful bites on man. This is the way biters possibly afford a colony some protection against squirrels that are known to eat the contents of aphid galls (Gailer, 1956; Aoki *et al.*, 1977; Aoki, 1979).

The proportion of biters or soldiers in a colony, and whether they are sterile, possibly depend on the probability of attack by natural enemies and the constraints on defence imposed by structure and feeding biology.

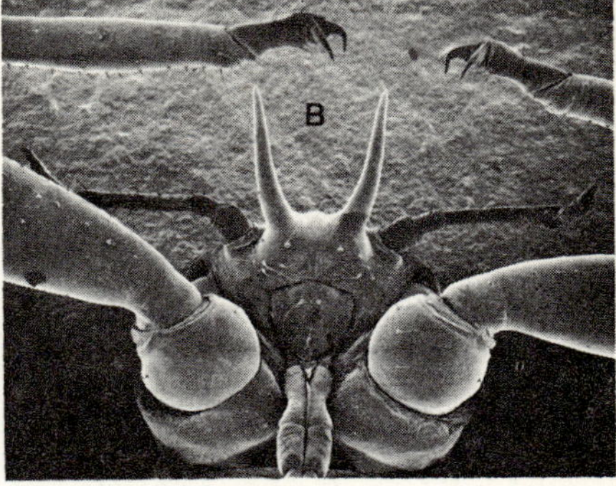

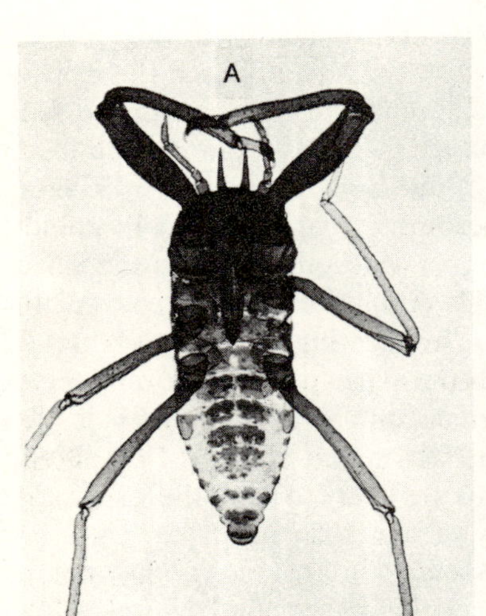

Figure 4.1 Soldiers of *Pseudoregma alexanderi*: *A* shows the enlarged prehensile forelimbs and frontal horns; *B*, a ventral view of a soldier's head; and *C*, a group of soldiers attacking a syrphid larva. (Photographs by Dr. Shigeyuki Aoki.)

Dispersal

As even trees eventually die or are killed, it is advantageous for aphids to disperse. Individuals of most morphs are capable of moving to alternative parts of their host plants or to adjacent plants. In some species, like the sycamore aphid, adults of most generations possess wings and are highly mobile as they retain the ability to fly throughout adult life. In species that show alary dimorphism, minor movement within and between adjacent plants is mainly undertaken by apterae, but alatae disperse over greater distances. Dispersal is most noticeable when aphids are abundant, especially in summer.

What constitutes a crowd?

In the vetch aphid (*Megoura viciae*), it is the tactile stimulation associated with crowding that induces the development of winged forms. When several apterous mothers that have been reared in isolation are confined together in a small glass tube for 24 hours and are then placed individually on bean plants, they give birth to alatae. Apterous mothers confined individually in glass vials for the same period give birth only to apterous offspring (Figure 4.2). This

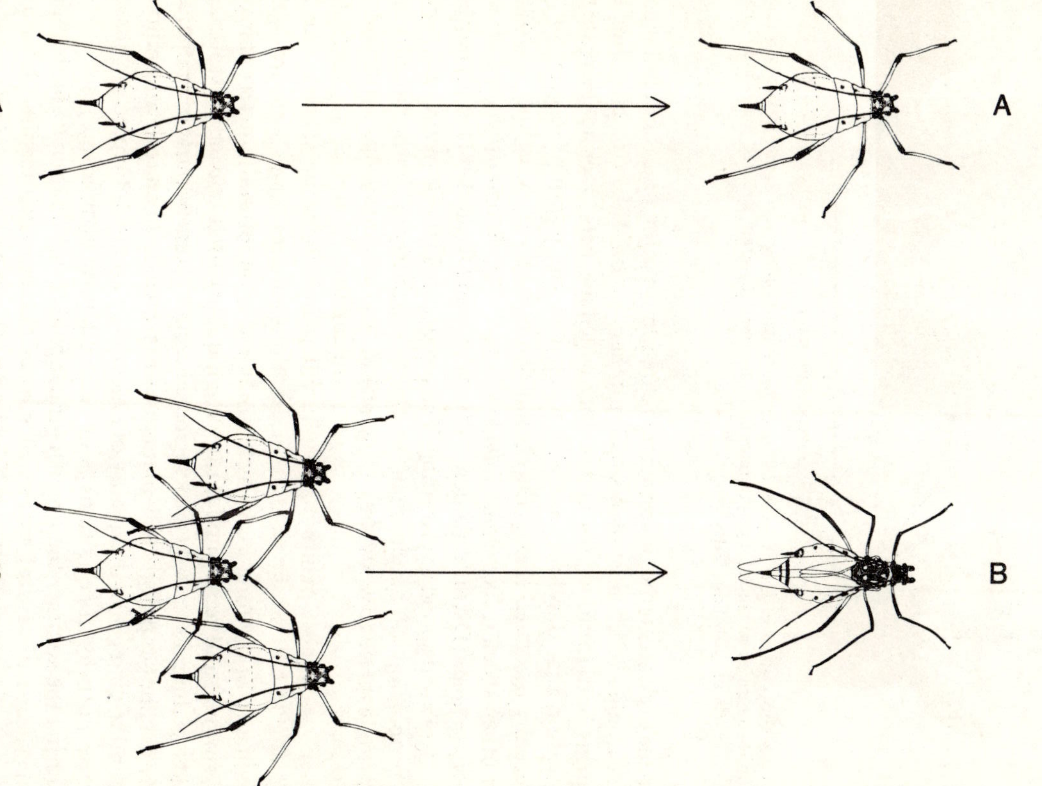

Figure 4.2 Diagram of the effect of rearing the vetch aphid (*Megoura viciae*) in isolation (*A*) or in crowds (*B*) on the production of alatae.

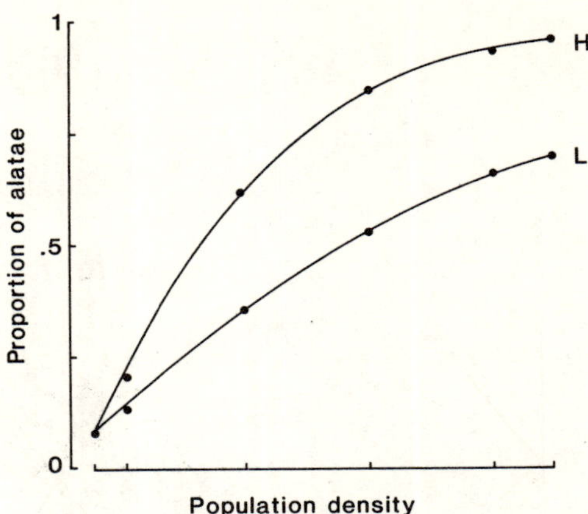

Figure 4.3 Proportion of the progeny of *Aphis fabae* developing into alatae in relation to the population densities at which they were reared, for mothers themselves reared under low (L) and high (H) levels of crowding. (After Shaw, 1970.)

happens even when the offspring are reared in isolation and is not influenced by the quality of the host plants their mothers fed on. Likewise, crowding the offspring born to either crowded or uncrowded mothers, does not increase the proportion that develops into alatae. That is, the response to crowding is maternally mediated (Lees, 1967), and two aphids can constitute a crowd.

In other species, like *Aphis fabae* (Shaw, 1970), *Rhopalosiphum padi* and *Sitobion avenae* (Dixon and Glen, 1971; Watt and Dixon, 1981), both mothers and offspring respond to crowding. The more intense the crowding, the higher the proportion of alatae that develops, especially if both mothers and offspring experience crowding (Figure 4.3).

Host quality

The quality of the food available to an aphid is important in determining its size, survival and reproductive rate. High-quality food is usually available to aphids feeding on actively growing or senescing plants or parts of plants, and often declines rapidly when plants cease growing. Therefore an ability to respond to changes in the quality of phloem sap and to anticipate the onset of adverse nutritional conditions is of great adaptive significance.

During the summer the English grain aphid (*Sitobion avenae*) feeds on the flower heads of several species of grasses that are rich feeding for only a relatively short period. Winged forms develop in response to both pre- and post-natal crowding and to changes in the nutritional quality of a host plant associated with the ripening of the seed (Figure 4.4). When mothers reared on the ears of grasses at flowering are crowded in glass vials for short periods, they give birth to proportionally fewer winged offspring than mothers

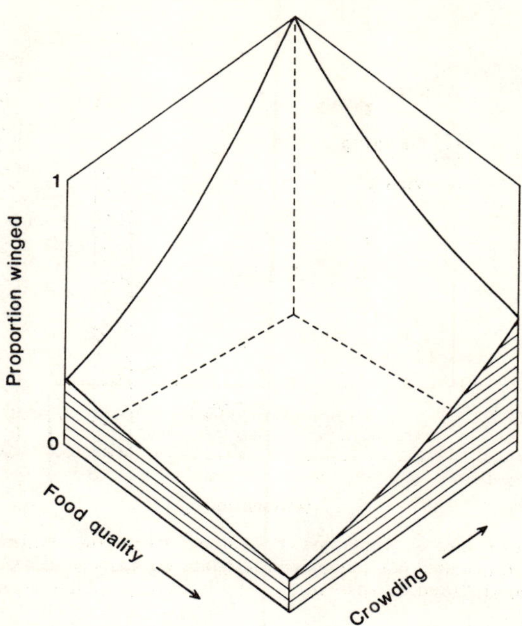

Figure 4.4 The proportion of the English grain aphid (*Sitobion avenae*) developing into alatae in relation to food quality and crowding. (After Watt and Dixon, 1981.)

similarly treated but reared on mature ears (Watt and Dixon, 1981). Similarly the pea aphid (*Acyrthosiphon pisum*) shows a heightened response to crowding when developing on mature leaves, and like the rosy apple aphid (*Dysaphis devecta*), can even produce alatae solely in response to changes in host quality (Sutherland, 1969*b*; Forrest, 1970).

Spring migrants of *R. padi* develop on the primary host, bird cherry, in response to both crowding and changes in host quality. When *R. padi* is reared in isolation on maturing bird cherry leaves, an increasing proportion develops into alatae in subsequent generations, but no alatae develop if *R. padi* is reared in isolation on young leaves (Figure 4.5). However, when the same species is crowded on young leaves it readily produces alatae (Dixon and Glen, 1971).

Rearing nymphs and adults of *Myzus persicae* in isolation on a synthetic diet has revealed that the composition of the food only partially controls wing polymorphism in this species (Sutherland and Mittler, 1971).

Although apterae tend to develop mainly on 'good' food, a high proportion of apterae develops on poor diets if, before giving birth, their mothers are allowed to walk on and superficially probe the surface of a nutritionally good host plant. In this case some feature of the plant surface is acting as a 'token stimulus' of good quality (Kunkel and Mittler, 1971). Likewise it is possible that non-nutritional components of an aphid's diet may also act as token

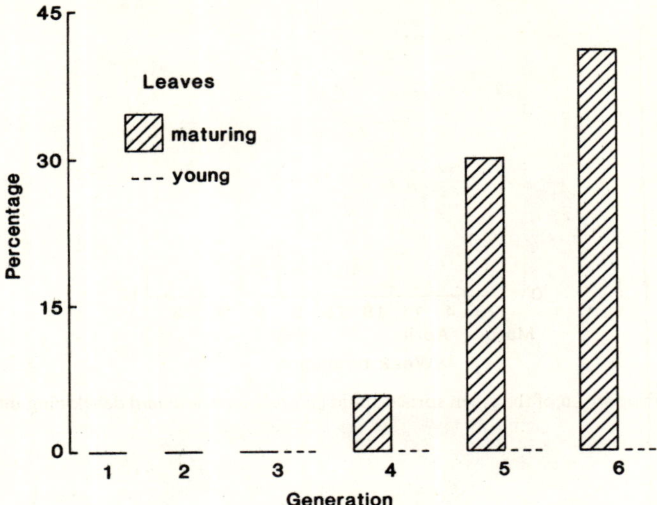

Figure 4.5 The percentage of emigrants developing when the bird-cherry oat aphid (*Rhopalosiphum padi*) is reared continuously in isolation for six generations, starting with the fundatrix, on maturing (▨) and young leaves (– – –) of bird cherry. (After Dixon and Glen, 1971.)

stimuli indicating changes in the host plant, and although an aphid's response is associated with changes in the quality of its food, perhaps it is not responding directly to that change.

However, on 'very poor' synthetic diets mainly apterae develop in *M. persicae* (Mittler and Kleinjan, 1970). This corroborates observations on other aphids developing on poor-quality hosts (Schaefers and Judge, 1971; Dixon, 1972*a*). Alatae generally take longer to reach maturity, are smaller and less fecund than apterae, especially in poor habitats, and if very small, they are poor fliers. Therefore in very harsh conditions it would be advantageous for a clone to develop apterae better able to survive until conditions improve, rather than allocate limited resources to dispersal.

Day-length

In several species alatae are produced at a definite time of year (Figure 4.6). This is particularly so for the alate gynoparae and sexuparae that give rise to the males and females in autumn (p. 69). However, alates also appear at other times. In the green spruce aphid (*Elatobium abietinum*) the maturation and flight of the alatae coincide with bud burst of spruce. Although the proportion of green spruce aphids that become alate is influenced by crowding, host quality and temperature, the dominant factor is the increase in day-length in spring. The maximum critical day-length is 14 hours, beyond which the proportion developing into alatae declines (Figure 4.6). This response is not

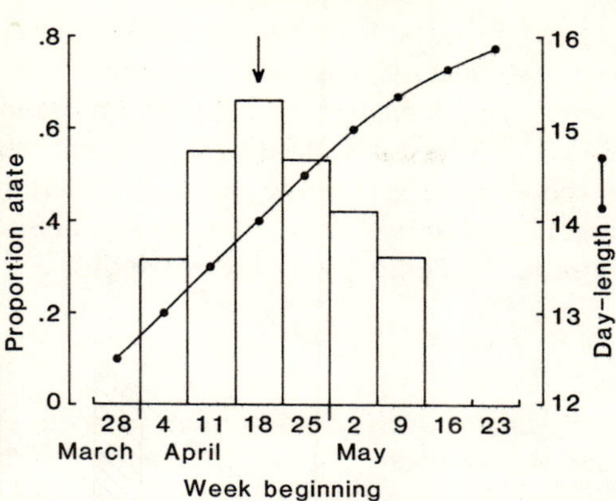

Figure 4.6 Proportion of the green spruce aphid (*Elatobium abietinum*) developing into alatae in relation to their time of birth and day-length (↓ = maximum critical photoperiod). (After Fisher, 1982.)

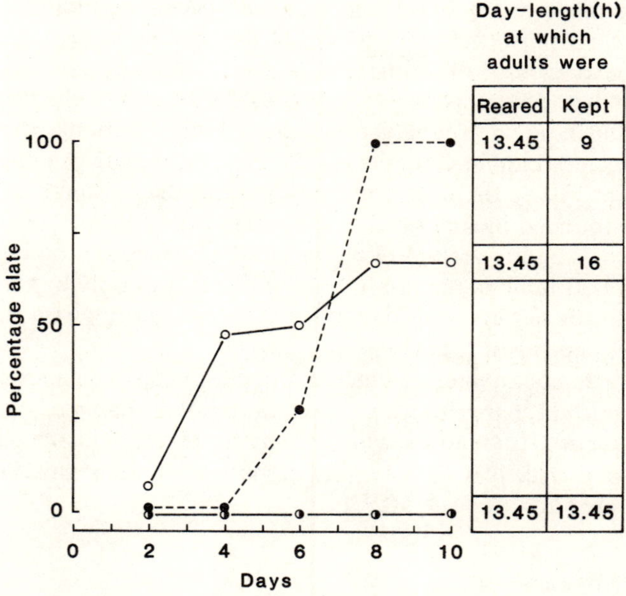

Figure 4.7 Percentage of alatae amongst the offspring produced every two days by *Myzus persicae* when reared and kept at the same and different day-lengths. (After Matsuka and Mittler, 1978.)

influenced by the generation of aphid, being just as marked and occurring at the same time, whether the aphid overwintered viviparously or as an egg (Fisher, 1982). Increased alatae production in response to changes in day-length has also been reported for *Myzus persicae* (Figure 4.7) (Matsuka and Mittler, 1978) and *Cinara todocola* (Yagamuchi, 1976).

Alate production and dispersal has been regarded as a means of regulating numbers, and in several species it is most marked when there are large populations. However, the production of alatae at a particular time of year, even when the aphids are reared in isolation, reveals that the primary role is dispersal. In many species there is possibly only a short period when dispersal is likely to be successful, and the response to changes in day-length enables the aphid to synchronize its development with this window.

Maternal control

It takes at least one generation to produce alate individuals. Therefore by responding to cues that indicate an impending deterioration in habitat quality, aphids can switch to producing alatae in advance of the adversity taking place. In species in which there is a maternal switch mechanism, it is possibly under neural control and occurs just as an embryo is released from an ovariole or is actually passing down an oviduct (Lees, 1979). That is, the 'decision' is made just before birth. In other species, commitment is made during the early postnatal development of the nymph, in response either to conditions experienced by both mother and nymph or to the conditions experienced by the nymphs only (Figure 4.8). As the rate of change in habitat quality is influenced by several factors and is therefore variable, then the later in development an aphid commits itself to switch to alate development, the more likely it will be that alates are produced at the optimum time.

Colonies founded by alatae initially contain only apterous offspring, and later an increasing proportion of alatae develop. There is a biological clock mechanism that inhibits the production of alatae, not only by the founding alate but also by her apterous daughters. This refractory phase is time-, not generation-, dependent (Lees, 1966).

The operation of an interval timer and the response to increased crowding result in a gradual change to alate production, rather than a dramatic switch after a certain time has elapsed. This presumably serves to increase the chance of survival and is likely to occur earlier in species that are regularly subject to heavy predator and parasite attack.

Multiplicity of cues

Although *Megoura viciae* only produces alatae in response to tactile stimuli (Lees, 1967), this is not so for all species. *Myzus persicae* responds mainly to food quality and tactile stimulation, with day-length, temperature and non-nutritional plant factors enhancing or suppressing the response (Matsuka and Mittler, 1978). By responding to a number of stimuli rather than one, aphids can possibly achieve a closer and more reliable tracking of environmental conditions. This enables them to produce the more fecund and faster developing apterae while a host plant is favourable, and when it becomes

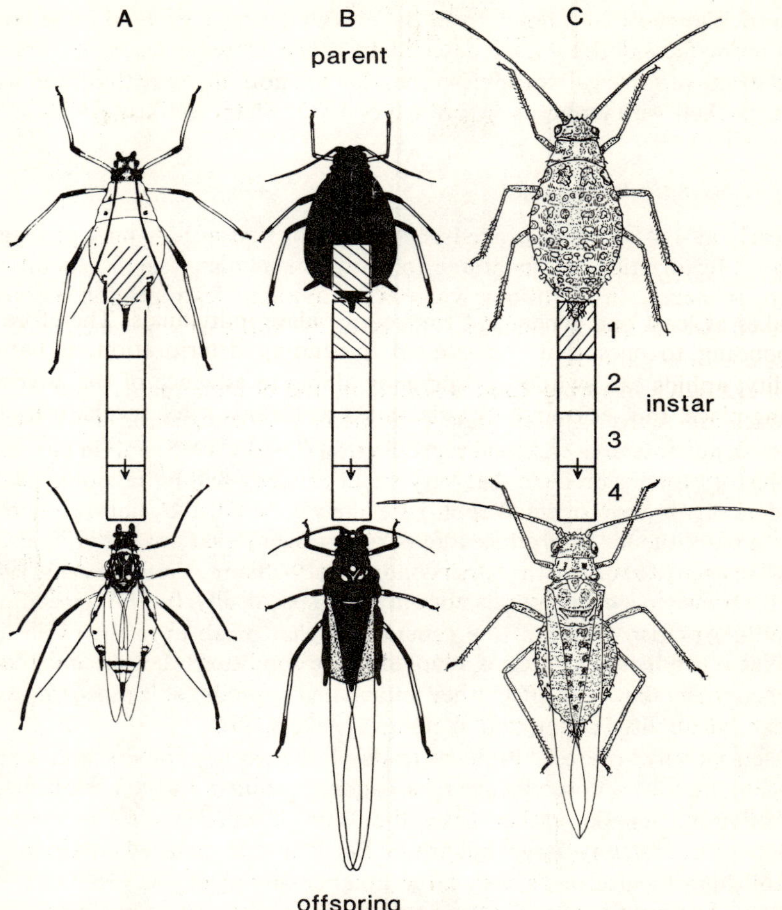

Figure 4.8 Periods during development when aphids are induced to become alate (▧). *A, Megoura viciae*—prenatal, *B. Aphis fabae*—pre and post-natal, and *C. Therioaphis maculata*—postnatal. (After Lees, 1967; Toba *et al.*, 1967; Shaw, 1970).

unfavourable to switch gradually to producing alatae that disperse to other plants.

Reproduction

Aphids are noted for their great reproductive potential. If all the offspring survive then in the course of a season the descendants of one female 'contain more substance than 500 000 000 stout men' (Huxley, 1858). Some aphids produce hundreds of offspring. However, certain morphs, for example oviparae of the Pemphiginae, produce only one. Early generations of apterae, in particular those that hatch from the overwintering eggs, tend to be larger and more fecund than their descendants. The extreme examples are the fundatrices of some gall forming species that are nothing more than 'awkward bags' packed with embryos (Mordwilko, 1935; Shaposhnikov, 1977).

Parthenogenetic reproduction

As early as 1745 Bonnet proved beyond doubt that aphids may propagate without fertilization and continue to do so for as many as 10 generations. Then, under certain conditions, winged or wingless males appear and copulate with wingless oviparous females. This is the cyclical parthenogenesis characteristic of most aphids.

At egg hatch in spring, food is abundant and of high quality, and natural enemies are scarce. Under these conditions parthenogenesis has a strong selective advantage over sexual reproduction (Dixon, 1985) and in particular life-history theory predicts that very small progeny will be favoured. When food quality is poor, or small aphids are likely to be killed by natural enemies, then a parent tends to produce a few large progeny that are better able than small progeny to survive the harsh conditions (Williams, 1966; Stearns, 1976). This tendency occurs in aphids not only interspecifically, but intraspecifically as well. Aphids of each of the generations that make up a life cycle have specific reproductive strategies adapted to the conditions they are most likely to encounter; that is, in effect, they anticipate the predictable seasonal trends in habitat quality (Dixon and Wellings, 1982; p. 79).

Alternatively, the evolutionary trade-off between numbers and size of progeny may be a consequence of a packing problem in the abdomen of a reproductive female, rather than the lack of nutrients to support the production of many large offspring. Early in the year when clones are establishing themselves, a very large number of progeny is produced, very rapidly. The multiplicity of offspring can spread the risk of misfortune, and improve the chances of finding good hosts and of survival of the clone, which may be more important than the size of the individuals. Rate of production is determined by the number of egg tubes (ovarioles) an aphid has, so there would be an advantage in increasing the number of ovarioles in the early generations. However, in order to maintain continuous production each ovariole contains a number of embryos in different stages of development. This could lead to a serious packing problem that the globular shape of some fundatrices may allow for. The mother's need to remain agile to avoid natural enemies and seek out new feeding sites limits the volume of embryos, except, notably, in gall-forming species, and these are afforded protection and good feeding conditions by the gall.

Many more ovarioles could be accommodated within an aphid's abdomen if the size of the embryos they contained was reduced (Figure 4.9). The tendency of one morph relative to another to increase progeny size with a

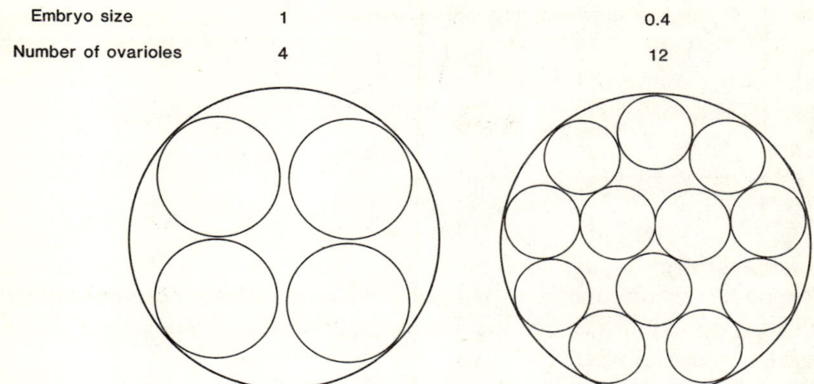

Figure 4.9 Diagram of a cross-section of the abdomen of an aphid showing the relationship between embryo size and number of ovarioles.

decrease in average ovariole number (Dixon and Dharma, 1980*b*) appears to support this idea. However, although the reproductive rate and ovariole number within a morph are positively correlated (Wellings *et al.*, 1980), there is no correlation between ovariole number and progeny size at this level (Ward *et al.*, 1983). Therefore although limitations on the space available in which to mature embryos may be a factor determining progeny size in some fundatrices, it is not apparently a factor influencing individuals of later generations, in which partitioning of limited nutrient resources between embryos and reserves is more important (p. 79).

Whether an aphid has or lacks wings affects its reproductive rate. When the viviparous generations are all winged, as in the sycamore aphid (*Drepanosiphum platanoidis*), the individuals usually have relatively low reproductive rates, compared with the wingless viviparous individuals of more polymorphic species. Alatae of aphids that do show alary dimorphism take from 8 to 26% longer than apterae to reach maturity (Table 4.1), and although they may initially have a reproductive rate comparable to apterae (e.g. *A. fabae*; Dixon and Wratten, 1971), they are more likely to have a lower overall reproductive rate (Wratten, 1977). Alates also give birth to smaller offspring than apterae of the same species do, which may be an adaptation of alates to the generally better quality of the plants they have colonized, than they have left (p. 89). However, if alatae produced offspring as large as apterae do, the reproductive rate of the alatae would be even further reduced if the same total allocation of resources was given to reproduction. Thus in order to colonize a new habitat the rapid production of fewer offspring may be more important than their size at birth, and this could also have contributed to the evolution of small size of offspring of alates.

Winged individuals of species that show alary dimorphism autolyse their wing muscles shortly after settling on a plant (p. 89). This prevents further

Table 4.1 Duration of development of apterae and alatae

Aphid	Temperature °C	Duration of development in days: Apterae	Alatae	Percentage increase	Source
Elatobium abietinum	15	15.6	18.0	15	Fisher, 1982
Metopolophium dirhodum	15	13.2	14 3	8	Thornback, 1983
Rhopalosiphum maidis	10	21.4	26.9	26	Noda, 1960
	25	5.0	6.0	20	
Rhopalosiphum prunifoliae (insertum?)	10	21.3	25.7	21	Noda, 1960
	25	4.7	5.5	17	
Sitobion avenae	10	21.4	26.9	26	Noda, 1960
	25	5.0	6.0	20	

dispersal and possibly has the advantage of making a small amount of protein available for embryo development (Dixon, 1971). However, the brachypterous alate of *Drepanosiphum dixoni*, which differs from the macropterous morph in lacking indirect muscles and not being able to fly, is 32% more fecund (Dixon, 1972*b*). Therefore the maintenance of wing muscles is costly, and by autolysing or not developing them, aphids free resources for the development of offspring.

The shorter developmental time and greater reproductive rate of apterae, compared with alates, also gives them a higher rate of increase. A first step in the evolution of apterousness was possibly a simple increase in fecundity with the appearance of brachypterous forms or as a result of wing muscle autolysis. Later apterous development could then give the additional advantage of a shorter developmental time (Table 4.1) and lower temperature thresholds for development (Noda, 1960).

Sexual reproduction

Continued parthenogenetic reproduction was thought to result in aphids becoming effete. Sexual reproduction supposedly invigorated a clone, and the 'spermatic force', which weakened in its passage between successive generations of cells, determined the number of generations and proliferation of a clone. Huxley (1858) was the first to show that if aphids were kept warm and supplied with food they could reproduce parthenogenetically without deterioration, apparently indefinitely. This is also supported by the existence of many anholocyclic species of aphids. It was not until 1924 that Marcovitch demonstrated the role of photoperiod in the induction of sexual forms. This was shortly after the discovery of the effect of photoperiod on flowering in plants. The role of sexual reproduction in producing genetic diversity is considered in Chapter 5.

Extrinsic factors

At the onset of autumn in temperate regions many species of aphids switch to the production of sexual forms; with each clone (evolutionary individual) producing both egg-laying females (oviparae) and males, i.e., a clone is hermaphrodite. Although associated with short days, the switch is actually triggered by long nights (Figure 4.10) (Lees, 1973).

However, aphids live in a multivariate environment in which the time of leaf fall and the coolness of the autumn vary from year to year. Parthenogenetic reproduction continues as late as is compatible with successful egg laying. Optimally, sexual forms are produced only when the interval before leaf-fall is equal to the time required for development, migration, mating and oviposition. As aphids are poikilothermic they need longer to complete the transition in cooler autumns. It is also likely that the average temperature in autumn affects the time of leaf fall.

The mean temperature in July can be used as a positive predictor of the mean temperature in August (Ward *et al.*, 1984). Therefore by producing sexuals at shorter and shorter night lengths at lower temperatures (Figure 4.11), aphids can initiate sexual development earlier in the years when there is a high likelihood of a cool autumn.

Field results for several species indicate that the initiation of sexual forms is advanced more than is necessary to compensate for the extra time needed for development at the lower temperatures. It is possible that plants also respond to predictors of an early onset to winter, as sometimes they shed their leaves

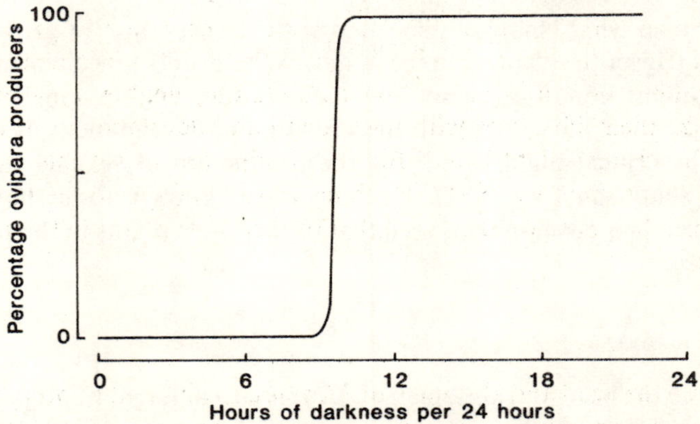

Figure 4.10 The percentage of vetch aphids (*Megoura viciae*) producing sexual females (oviparae) in relation to the numbers of hours of darkness per day they were reared under. (After Lees, 1959).

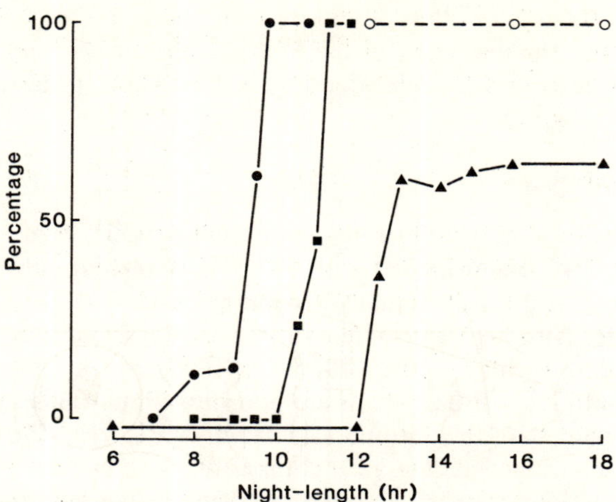

Figure 4.11 Percentage of the bird cherry-oat aphid (*Rhopalosiphum padi*) developing into gynoparae and males in relation to the night-length they are reared under, at 10° C(● —— ●), 14° C (■—— ■), and 18° C (▲—— ▲, ○ – – – ○ 3rd generation). (After Dixon and Glen, 1971.)

early. Thus the early initiation of sexual forms in cool autumns possibly compensates for both the extra time necessary for development as well as an earlier leaf fall (Ward *et al.*, 1984).

In several aphid species sexual production occurs in response to changes in the host plants. In temperate regions the short-day conditions in autumn induce plants to cease growth and become dormant. Aphids living in continuous darkness, at relatively constant temperatures, and often at considerable depths on the roots of these plants, still produce sexual morphs as the plants become dormant. *Aphis farinosa* and *Dysaphis devecta* live above ground, but produce sexual morphs when night-length is at its shortest, responding to the cessation of shoot growth of their host plant (Forrest, 1970). It is unknown what changes associated with the cessation of growth of the host plant trigger the changes in the aphids. Whatever the mechanism, it does, like long-night conditions and low temperature, enable some aphids to synchronize their life cycles with the growth and development of their host plants. The critical night-length for the production of sexuals varies with latitude (Shaposhnikov, 1981). Nothing is yet known about how aphids synchronize their development with that of their host plants in the subtropics and tropics.

Maternal control

By exposing the head and abdomen of *Megoura* to different photoperiods and destroying groups of cells in the brain by microcautery, Lees (1964) and Steel and Lees (1977) have revealed that neurosecretory cells in the pars intercerebralis, and tissue lateral to them, control sex determination. The light

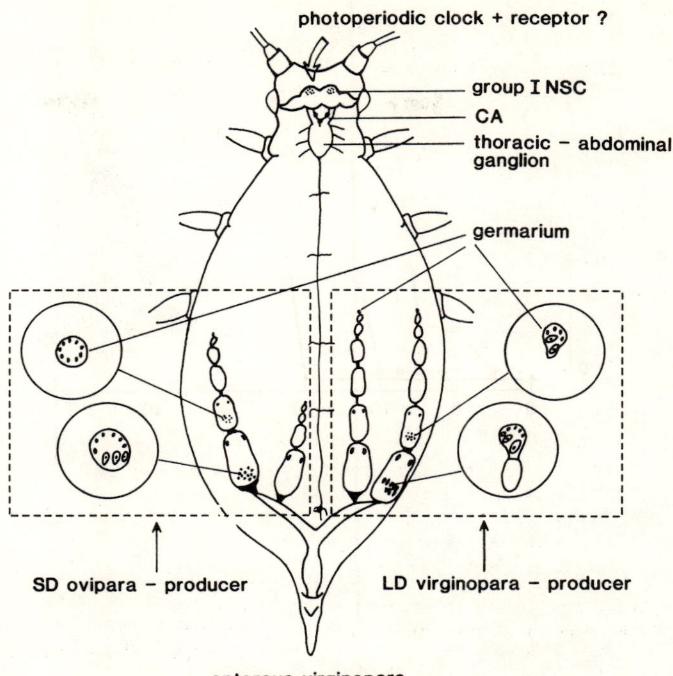

Figure 4.12 Representation of the adult apterous virginopara of *Megoura viciae*, showing some of the internal structures. The ovarioles contain chains of embryos the largest of which under long-day conditions (right) contain functional germaria, which have already produced one or two embryos (encircled), and under short-day conditions (left) the embryos are destined to become oviparae and the germaria produce yolky haploid eggs. (NSC, neurosecretary cells, CA, corpus allatum). (After Lees and Hardie, 1981.)

receptor and probably the photoperiodic clock are located in the lateral area, and the neurosecretory cells are the endocrine effectors producing a virginopara determinant. This determinant may be delivered direct to the embryos via the thoracic ganglion or released into the blood as a neurohormone (Figure 4.12). However, in the gynoparae of *A. fabae* the photoperiodic clock controls nymphal development through the corpus allatum which produces the juvenile hormone that influences both the external form and the type of gonads of the nymph. The amount of juvenile hormone is also thought to affect the proportion of males produced by apterous virginoparae of *Myzus persicae* kept under long-night conditions (Lees and Hardie, 1981; Hales and Mittler, 1983).

Aphid clocks

Although many aphids hatch from eggs early in the year and are present in the spring when night-length is as long as in autumn, the sexuals rarely appear in

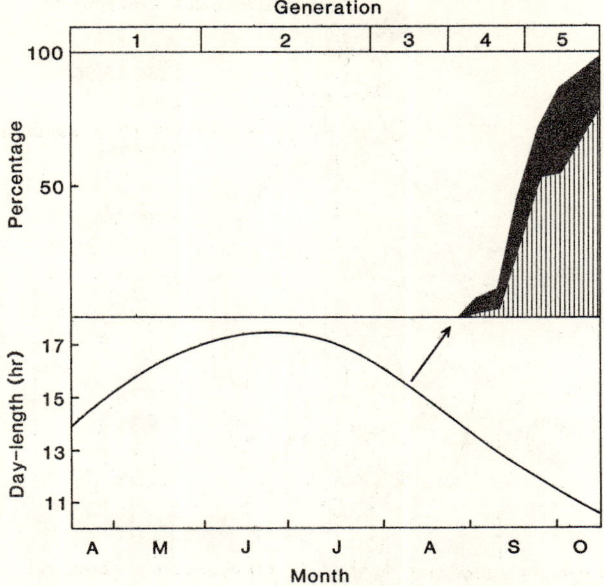

Figure 4.13 The relationship between day-length and the appearance in the field of sexual morphs of the sycamore aphid (*Drepanosiphum platanoidis*). □, viviparae; ■, males; ▥, oviparae. (After Dixon, 1971*a*.)

the field before autumn (Figure 4.13). This is because of the operation of an intrinsic timing mechanism referred to as a 'facteur fondatrice' (Bonnemaison, 1951) or as an 'interval time' (Lees, 1960, 1966).

In *Megoura* no oviparae or males are produced by young clones founded by fundatrices for 80 to 90 days, even if they are kept in a constant long-night regime at 15° C. This response is independent of generation, as it has been shown that a lineage of the early-born mothers of each generation will pass through four generations before ovipara producers develop, whereas only two generations are required in a parallel lineage derived from the last-born (Figure 4.14). Although the timing mechanism(s) is temperature-dependent, the slight variations in the time interval are unimportant provided the period of delay extends into early summer when the short night-length will take over the role of the inhibitor.

However, in the lime and sycamore aphids the interval timer mechanism does not function independently of the environment while it is 'running out'. With the passage of time the ratio of the number of oviparae to virginoparae produced by aphids in successive generations gradually increases, but does so more rapidly in short, than in long, photoperiods (Figure 4.15). Incomplete inhibition of the interval timer in autumn permits part of a sycamore aphid population to continue reproducing parthenogenetically in the short days

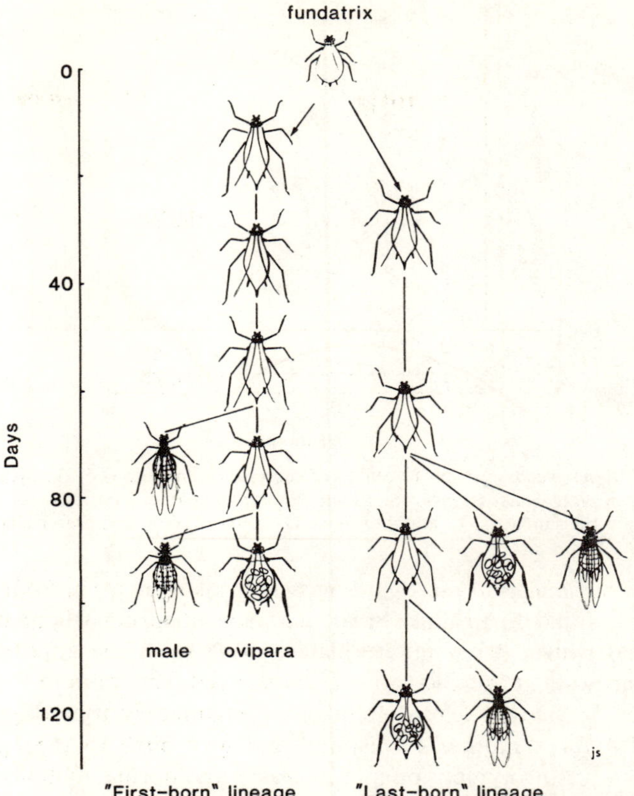

Figure 4.14 A trans-generation interval timer in *Megoura viciae*. Two lineages from one fundatrix were exposed permanently to a short-day regime. The appearance of males (winged) and oviparae (with eggs) is delayed from 75 to 90 days irrespective of the generation number. (After Lees, 1979.)

when the senescent foliage of sycamore provides a rich source of food. This is advantageous in those autumns when leaf fall is late (Dixon, 1971*a*, 1972*c*).

Survival

In summer many plants, especially shrubs and trees, cease growing and, until their leaves become senescent at the onset of autumn, are poor hosts for some aphids. Host-alternating aphids seek out and colonize plants that are still growing during this period, whereas several autoecious species of aphids aestivate. The sycamore aphid anticipates the onset of adverse conditions associated with the cessation of growth of its host; its summer morph matures with poorly developed gonads and a well developed fat body and may not reproduce for up to 8 weeks (p. 79). Another aphid also living on sycamore,

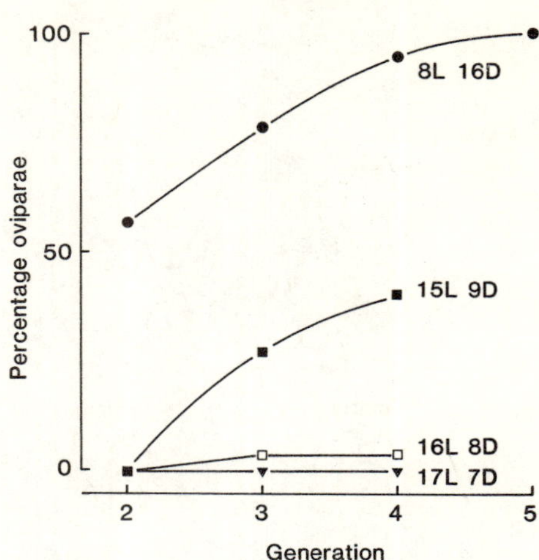

Figure 4.15 Progressive changes in the ability of successive generations of the sycamore aphid (*Drepanosiphum platanoidis*) to produce sexual females (oviparae) in a range of day-lengths. Generation 1 is the fundatrix, L, hours of light, D, hours of darkness. (After Dixon, 1971*a*.)

Periphyllus testudinaceus, aestivates as a first instar nymph. Poor nutrition induces this aphid to produce small flattened offspring whose bodies are covered and fringed with minute plates giving them the appearance of a tortoise, and so their specific name. If transferred from mature to young, or from young to mature leaves, it takes approximately four days at 15° C (Figure 4.16) for a mother to respond and give birth to the appropriate nymphs: aestivating nymphs on mature leaves and normal nymphs on young leaves. Aestivating nymphs attach themselves very closely to the leaves and are very difficult to remove, by predators or wind brushing the leaves together. The short photoperiods of autumn induce the aphid to resume growing and develop into sexuparae that give birth to males and oviparae (Bonnemaison, 1956; Shearer, 1976). *Thelaxes dryophila,* living on oak, has a similar life cycle, and the alatae that are produced when the leaves mature give birth to minute nymphs that conceal themselves at forks in leaf veins. They remain there until the short photoperiods in autumn induce them to develop directly into very small neotenous males and females.

Aestivation is not only to be found in tree-dwelling aphids. Autoecious species living on herbaceous plants may also have aestivating forms as is well illustrated by *Aphis urticata* living on nettles. The small yellow form that lives spaced out on the mature leaves of nettle in summer is very different from the ant-attended dark-green aphid found clustered on the growing points of nettles in spring and early summer. They are so different that for some time the summer form was placed in a different genus and species, *Pergandeida stanilandi.*

Many species can overwinter viviparously instead of in the egg stage. In

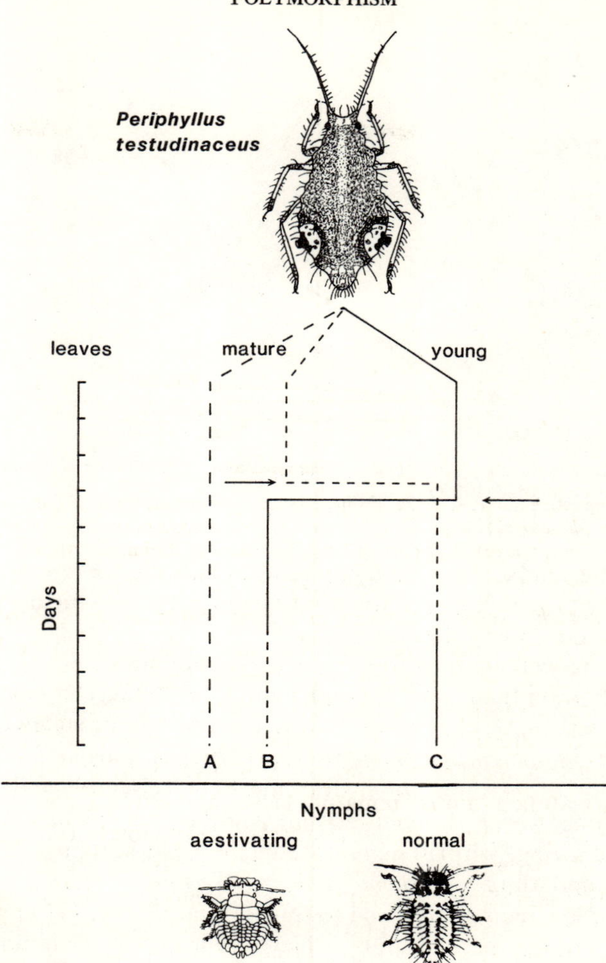

Figure 4.16 Production of aestivating (broken line) and normal (solid line) offspring by apterous fundatrigeniae of *Periphyllus testudinaceus* when (*A*) kept as adults on mature leaves for 10 days, (*B*) kept on young leaves for 3 days and then transferred (←) to mature leaves, or (*C*) kept on mature leaves for 3 days and then transferred (←) to young leaves. (After Shearer, 1976.)

some a special overwintering form, called a hiemalis, is induced to develop in response to environmental cues. About a fifth of the autumn apterae of the lettuce root aphid (*Pemphigus bursarius*) can survive without a host plant for 48 weeks at 3° C (Dunn, 1959). They have poorly-developed gonads and a well-developed fat body (Judge, 1967; Sutherland, 1968). The environmental cue that induces the development of the hiemalis in this root feeding aphid is low temperature; at 10° C, 95% develop into hiemalis that overwinter in the soil, and the remainder develop into sexuparae that return to the primary host,

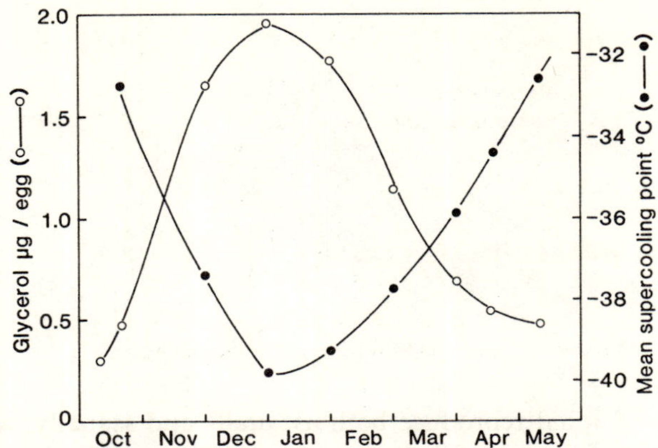

Figure 4.17 Changes during winter in the glycerol content and supercooling point of eggs of *Aphis pomi*. (After Sömme, 1969.)

poplar. At this temperature the hibernating aphids do not reproduce for 103 days, which is long enough for low winter temperatures to further delay their development until the following spring.

Eggs too are well adapted to survive periods of stress. Some can survive temperatures lower than −40° C (James and Luff, 1982). This cold-hardiness of eggs (Sömme, 1969), like that of the active overwintering stages of the green spruce aphid (*Elatobium abietinum*) (Parry, 1979*a*), is associated with changes in the levels of certain polyhydric alcohols (glycerol and mannitol) (Figure 4.17). However, the cold-resistance of the eggs of *Cinara pilicornis* is not related to their glycerol or mannitol content. Cold-hardiness of the eggs of this species and those of *Rhopalosiphum insertum* is related to previous experience of low temperature and to time of year (Parry, 1979*b*; James and Luff, 1982), but the mechanism by which it is achieved is unknown.

Division of labour, increasing the chances of survival of a clone, may also occur within a morph. For example in the gall-forming aphid *Pemphigus dorocola*, the first instar fundatrigenia have a cleaning role. They push droplets of honeydew and exuviae out through the small opening in the gall (Aoki, 1980). However, it is not recorded whether the first instar has special morphological adaptations for carrying out this cleaning function.

Summarizing, their clonal structure has enabled aphids to develop a division of labour in which each morph is adapted preferentially to either defence, dispersal, reproduction or survival. This is particularly true in the host-alternating Aphidinae and Pemphiginae. They track the seasons, producing the various morphs in response to a multiplicity of environmental cues and to changes in their internal clocks.

5 Cyclical parthenogenesis

Most species of aphid reproduce both asexually and sexually, with several generations of parthenogenesis between each bout of sexual reproduction. The production of sexual morphs is triggered by environmental cues (p. 51) and usually has an annual periodicity. Parthenogenesis is widespread in plants but in animals is commonly found only in rotifers, cladocera, cynipids and aphids. The earliest aphids are likely to have been oviparous, like the closely related Adelgidae and Phylloxeridae, with parthenogenesis and viviparity evolving as early as the Palaeozoic (p. 3). Although parthenogenesis is widespread in animals and plants it is relatively rare (Maynard Smith, 1984) but predominant in climax situations that are maintained by disturbance, which tends to temporarily denude an area of its animals and plants. The high rate of increase of parthenogens and the ability of an individual to establish a colony gives them considerable advantage over sexual species when colonizing new habitats (Cuellar, 1977). That all species of aphids have some parthenogenetic generations favours the suggestion that irrespective of whether they live on herbaceous plants or trees aphids are opportunists exploiting ephemeral habitats, which are continually being created (Dixon, 1985*b*).

As a consequence of parthenogenesis each genotype is represented by many individuals. As the individuals of a clone are likely to be widely dispersed within a habitat a clone is much less likely to be wiped out by natural enemies or to fail to locate a suitable host plant. Competition between clones for resources is occasionally likely to be intense. This, together with the variation in the ability of clones to locate and feed on certain plants, avoid death from natural enemies and withstand periods of stress determines which genotypes survive.

Both Huxley (1858) and Janzen (1977) viewed the increase in numbers by asexual reproduction as a process of growth, and Janzen suggested that a clone should be regarded as an 'evolutionary individual' equivalent to the body of a sexually reproducing organism although split up into a number of physically separate units. However, Janzen (1977) was wrong to equate reproduction with sexuality, and growth with asexuality, because there is a crucial difference between 'reproducing' to make two aphids and growing to make one aphid twice as large. Reproduction involves passing through a single-cell stage, therefore a mutation present in the genetic material of that cell can affect development to cause reorganization of the body. Growth does not involve passing through a single cell stage, and therefore a mutation can only affect a small group of cells and cannot cause reorganization of the body (Dawkins, 1982).

Intra-clonal variation

Cognetti (1961) claimed that genetic recombination occurs during parthenogenesis and provides a significant and continued source of variation. Endomeiosis, as he called the process, could result in progeny differing genetically from their parents and the loss of heterozygosity. However, most of the histological evidence indicates that parthenogenetic eggs are produced by mitosis during which no chiasmata are formed (Suomalainen, 1950; Blackman, 1979; Suomalainen *et al.*, 1980). The stability of aphid clones heterozygous for colour (Muller, 1962) and enzymes (Blackman, 1979; Suomalainen *et al.*, 1980; Tomiuk and Wohrmann, 1981) also indicates that no *significant* amount of recombination occurs during the development of parthenogenetic eggs.

The experimental evidence presented by Cognetti and others in support of endomeiosis should be treated circumspectly as it involved selecting for features like apterousness, which is under environmental control (p. 38). In selection experiments in which environmental conditions were rigidly controlled, there was no significant change over periods of 18 to 170 generations in the percentage of individuals showing the character selected for (Ewing, 1916; Lees, 1966; Blackman, 1979).

Blackman (1979) attributes the undoubted varation that occurs in laboratory 'clones' to contamination, mutation, mitotic recombination and variability of gene expression. Contamination is important and the longer a clone is maintained the greater the risk. Mutations that result in a marked change in fitness, as, for example, resistance to an insecticide or ability to colonize another host plant, are likely to occur as frequently in parthenogenetic as in sexually producing populations. In addition, genetic changes involving dissociation or rearrangement of chromosomes may also occur and may even be quite common (Blackman and Takada, 1965). Mitotic recombination whereby crossing over can occur and heterozygous genes can segregate in somatic cells is a rare event and has only been conclusively demonstrated in some Diptera.

In a given lineage an aphid can be winged or unwinged, parthenogenetic or sexual, hence different genes are active at different times. The sensitivity to environmental cues can be determined by the genotype, for example, the different response of green and pink forms to crowding (Sutherland, 1969*a*) and geographical variation in the photoperiodic responses and developmental threshold in the pea aphid (Campbell *et al.*, 1974; Lamb and MacKay, 1983). Chance may also be important in determining whether an embryo develops into a particular morph, especially when conditions do not significantly favour one or other course of development, and may account for the mixed families observed under these conditions.

Thus there are several different ways in which variability could occur within aphid clones. Although genetic changes occur relatively rarely, the extremely large populations of many aphid species substantially increase the likelihood of such changes, and then parthenogenetic *reproduction* increases the likelihood of incorporating the changes at the organ level.

Adaptability of aphid clones

Blackman (1981) recounts that in 1953–54 a parthenogenetic female of the spotted alfalfa aphid (*Therioaphis trifolii* forma *maculata*), of Mediterranean origin, was introduced and established a clone in New Mexico. By 1956 it had spread through most of the southern United States. It quickly developed resistance to organophosphorous insecticides and in 1958 attacked a previously resistant variety of alfalfa.

Before 1960 the aphid overwintered as parthenogenetic females in the southern United States and migrated north each summer. Then, in the autumn of 1960, a population in the centre of Nebraska produced viable sexual morphs and passed the winter there in the egg stage. By 1964 the new variant had colonized Nebraska and extended the range of the species northward into South Dakota, an area previously not reached by the annual migration from the south, and another population had also started to reproduce sexually in Wisconsin. Today the aphid is found throughout the United States and reproduces sexually, overwintering as eggs over most of the northern part of its range. In the south-western states where sexual reproduction has not been recorded, clones capable of colonizing previously resistant cultivars of alfalfa continue to appear. Therefore, the lack of genetic recombination in parthenogenetic reproduction has not prevented this aphid adapting to new conditions (Blackman, 1981).

Aphid population size can be very large; for example, 2.5×10^6 aphids on a 20 m sycamore tree, 4×10^8 cereal aphids per hectare and 1.7×10^{11} aphids in one valley in California. With mutation rates of from 10^{-5} to 10^{-9} gene^{-1} generation^{-1} then even a rare mutation is likely to have occurred at least 1.7×10^2 times per generation in the Californian valley (Dickson, 1962). Therefore provided mutations occur, change becomes a matter of certainty rather than chance in such large populations. As an individual with an advantageous mutation may pass it to all its descendants, which can be as many as a hundred, the mutation is likely to spread very rapidly.

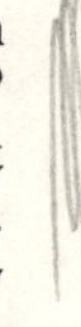

Genetic structure

Allozyme studies have revealed surprisingly little genetic polymorphism in aphids (Tomiuk and Wöhrmann, 1980). The apparent lack of genetic polymorphism may have resulted from studying enzyme systems that are basic to the functioning of all individuals within a population or be a consequence of competitive exclusion. However, the breeding of plants resistant to aphids and the study of colour polymorphism in aphids have confirmed that many aphid populations are genetically diverse.

At least four strains of the European raspberry aphid (*Amphorophora idaei*) differentially infest resistant varieties of raspberry. The four aphid strains result from segregation at two gene loci and the plant resistance to aphids from different combinations of 14 genes. Surprisingly, the aphid strain that can colonize most raspberry varieties is consistently the least common. This has been attributed to its low sexual fertility (Briggs, 1965; Knight and Alston, 1974). This gene-for-gene relationship between an aphid and its host plant indicates the great potential genetic variability that could exist in highly polyphagous aphids.

The pea aphid (*Acyrthosiphon pisum*) in north-west and central Europe is predominantly holocyclic and is made up of a number of strains that are each differently adapted to living on a range of species of leguminous plants (Müller, 1962). The green strains of this aphid introduced into North America, and more recently Australasia, lack the red allele common in European populations. The green strains are particularly well adapted to living on peas and have been assigned to the subspecies *A. pisum destructor*. Similarly, only a portion of the Old World *Therioaphis trifolii* genome was introduced into North America (p. 59).

Several species of aphid have stable colour morphs that are genetically determined: for example, the red allele is dominant to the green in the pea aphid (Müller, 1962) and in *Myzus persicae* (Takada, 1981). How do these strains manage to coexist? In certain polyphagous aphids, the individuals of one colour are better adapted to living on certain plant species within the aphid's host range, as in the pea aphid (Müller 1962; Markkula, 1963; Ueda and Takada, 1977). In *Myzus persicae* red anholocyclic strains (which do not reproduce sexually) are better able to survive winter conditions than the green strains; then in the spring the green strains become predominant, increasing in abundance relative to the red (Figure 5.1). Thus in addition to having different host ranges the various strains may be at a selective advantage at certain times of the year. However, the significance of the colour (brown, yellow, red, etc.) to each of the strains of these aphids is unknown. There is also evidence for density-dependent differences in genotypic fitness. Changes in particular gene frequencies with increasing population density in the rose aphid (*Macrosiphum rosae*) indicate that the reproductive rate of one genotype decreases, and another increases, when population density itself increases (Tomiuk and Wöhrmann, 1981).

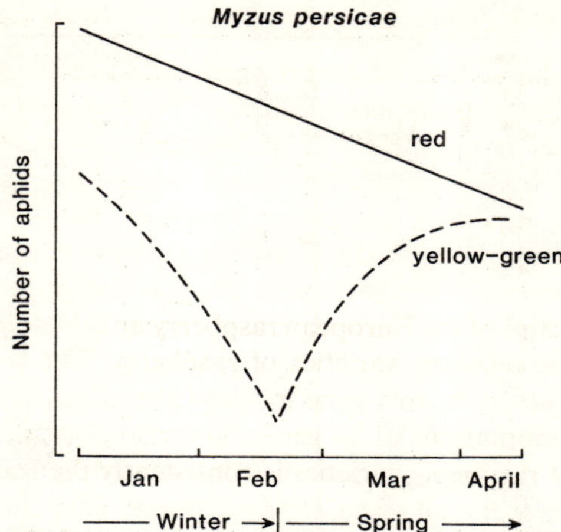

Figure 5.1 Population trends of red and yellow-green strains of *Myzus persicae* on Chinese mustard in the field from January through to April in 1978. (After Ueda and Takada, 1977.)

The range of genotypes present in aphid populations makes the breeding of aphid-resistant plants particularly difficult. In screening for resistant varieties of plants as great a range as possible of the naturally occurring aphid genotypes should be used. A rare resistance-breaking genotype could increase only too rapidly after the introduction and widespread planting of a new 'resistant' variety of crop plant.

Some species tend to become anholocyclic. A genetic change results in failure to produce sexual forms, and may result in the coexistence of holocyclic and anholocyclic forms of the same species, as in *Myzus persicae*. In response to short photoperiods, some clones produce oviparae and males (holocyclic); most produce some males and viviparous females (androcyclic) and a few do not produce sexual morphs at all (anholocyclic). These life-cycle traits are inherited and determined monofactorially, with androcycly recessive to holocycly. Androcyclic and anholocyclic clones are likely to be killed in severe winters. However, they can be generated afresh from the holocyclic clones (Figure 5.2) (Blackman, 1976). The proportion of holocyclic, androcyclic and anholocyclic strains in the population is likely to vary from year to year in response to winter conditions, which are a major factor in determining the relative fitness of the strains. When winters are consistently mild and the environment is uniform and predictable, it is likely that an aphid species will become anholocyclic.

Sex

Both sexual females and males are produced parthenogenetically in response to external and/or internal cues (p. 48). The sexual females have 2 X-

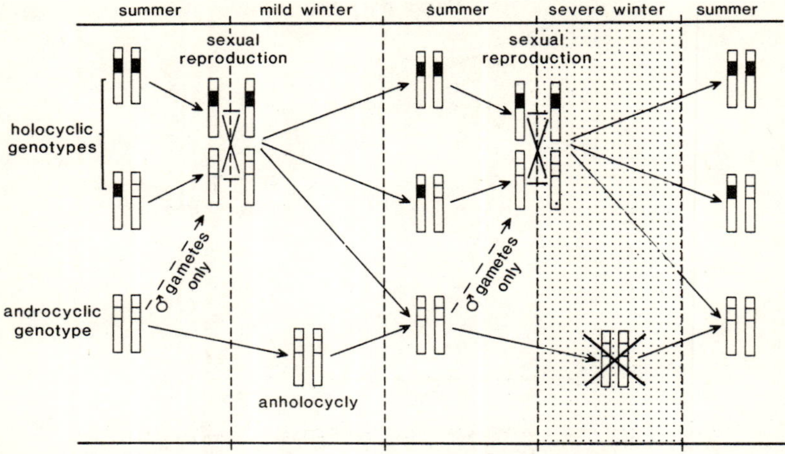

Figure 5.2 Maintenance of life cycle variation in *Myzus persicae* through three seasons. Androcyclic clones are generated afresh by the breeding system each year, even after a severe winter when parthenogenetic morphs outdoors are eliminated. (After Blackman, 1976.)

chromosomes as do the parthenogenetic females. In order to develop into a male, however, oocytes have to shed half the sex chromatin of the parent female in the course of their single maturation division.

The X-chromosomes undergo what Orlando (1974) refers to as a 'mini meiosis' in which they pair and then separate; the chromatids that remain on the equator of the spindle separate with the autosomes (Figure 5.3) and give rise to an XO male. Males are potentially capable of producing two types of sperm but during meiosis those without an X-chromosome (Figure 5.3) degenerate. As all viable sperm carry an X-chromosome fertilized eggs always give rise to females.

Why reproduce sexually?

As anholocyclic clones of aphids have been able to develop resistance to insecticides and colonize previously resistant cultivars of crop plants relatively quickly, and above all have a higher rate of increase than holocyclic strains, it is surprising that so few aphid species, only about 3%, are totally parthenogenetic. What is the adaptive significance of sex for aphids?

Aphids predominate in the temperate regions of the world and to overwinter in a cold-resistant resting stage, the fertilized egg, is an adaptation to temperate conditions. This only accounts for egg laying in temperate regions, not sex. Pseudosexual reproduction in which females produce eggs without fertilization occurs in the closely related Adelgidae and Phylloxeridae and several other groups of animals. Therefore, in addition to asking why most aphids reproduce sexually one can also ask why aphids combined

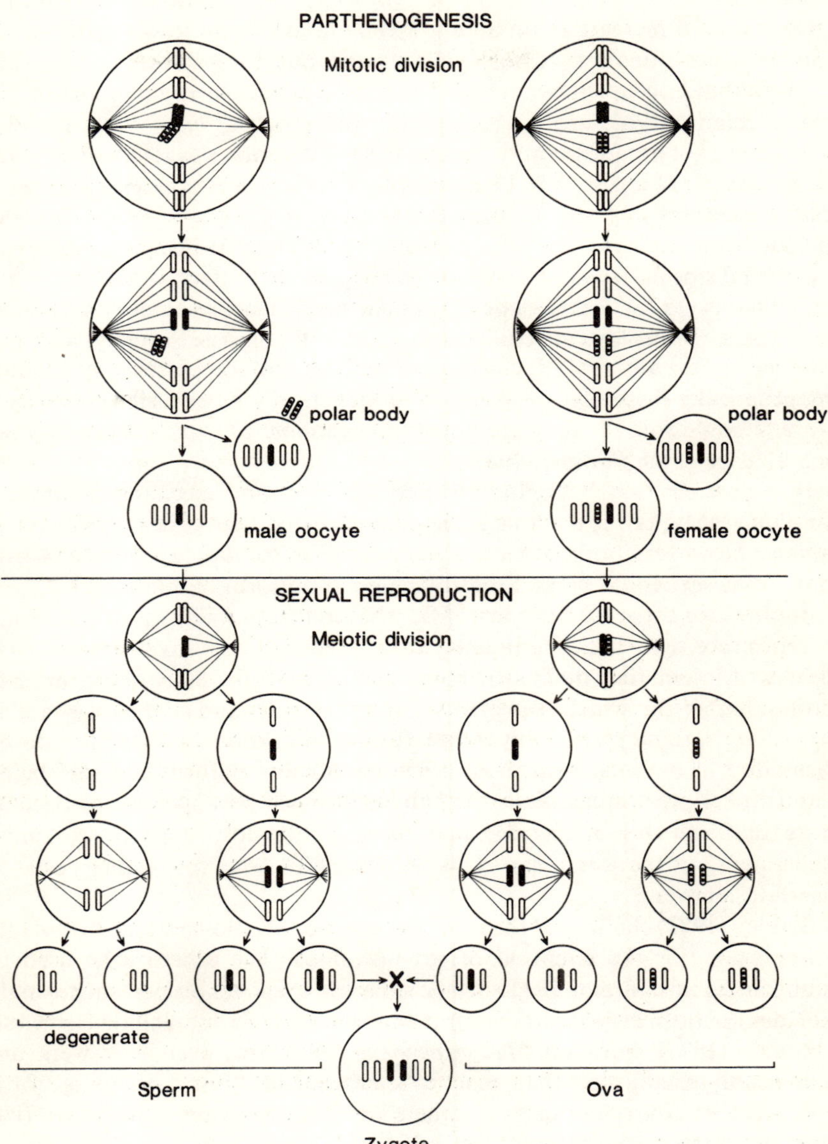

Figure 5.3 Diagram showing the behaviour of the chromosomes during the mitotic divisions leading to the development of male (XO) and female (XX) oocytes and the meiotic divisions in male and female aphids that result in the development of sperm and ova only from germ cells that carry an X-chromosome. (After Blackman, 1980.)

parthenogenesis with viviparity, and sexual reproduction with oviparity.

How can sex be maintained in the short term by selection? After all, the greater rate of increase of an anholocyclic mutant could lead rapidly to the competitive elimination of holocyclic strains, but it does not.

The habitat occupied by an aphid species is not uniform but consists of a spatial-temporal mosaic of many different patches, each with its own complement of organisms and resources. An individual's exact location often determines its fitness (p. 84). Thus one possible short-term advantage of sex is that it generates sibs with a range of genotypes, and a range is more likely to include the fittest genotype for a particular patch than the single genotype of an asexual sibship would do. This is the basis of the 'lottery model' proposed by Williams (1975). The genetically similar individuals of an asexual sibship have the same ecological requirements and will compete severely with one another. A sexually reproducing parent reduces this competition by producing genetically dissimilar offspring. Thus genetically diverse sibs could have more 'elbow-room' as they are potentially capable of exploiting more than one kind of patch (Young, 1981).

It is possible that both 'lottery' and 'elbow-room' mechanisms operate simultaneously, but it remains to be demonstrated that either is relevant to aphids. However, both models assume environmental heterogeneity and imply that sex would be at a disadvantage in uniform habitats.

Support for these models comes from the tendency of many pest aphids not to reproduce sexually, even in temperate climates. Crop environments have far fewer interacting biological components, and are more uniform than natural habitats in which host plants, natural enemies and competitors are all generating variability. Possibly the single most important factor is the reduced variability of the host plant with a few genetically uniform cultivars being planted over great areas. Thus the habitat of many pest species has become more uniform and predictable and there is less advantage to continual adjustment of the genotype. This is especially true for aphid pests of glasshouse crops.

Blakley (1982) challenged the idea that genetic variation among host plants is important for the retention of sex in aphids. The criterion he used to establish the existence of 'host-specific superior aphid genotypes' surprisingly excludes the raspberry aphid (p. 60) and his survey does not include Ueda and Takada's (1977) work on *Myzus persicae*. However, even if it were unequivocably established that genetic heterogeneity among plants is not a major factor in the retention of sex, there are other environmental factors that generate habitat heterogeneity for aphids.

In parthenogenetic strains, deleterious or non-functional alleles will accumulate, by what H. J. Muller (1964) has likened to a ratchet mechanism. Individuals with many mutant alleles will be selected against, but there is no way in which the overall load of deleterious genes can be reduced. Eventually, the size of the load will severely limit the possibility for further genetic change. However, the genetic recombination that occurs in sexual reproduction can expose the recessive deleterious mutants to selection in the homozygous form. Therefore, although a parthenogenetic clone may respond adequately to environmental change in the short term (p. 59), its long-term evolutionary prospects are bleak because it has no means of effectively exposing recessive deleterious mutations to selection. Because of the two-fold reproductive advantage of parthenogens over sexually reproducing species, however, it is difficult to account for the maintenance of sex by individual selection (Maynard Smith, 1984).

Timing of sex

Sexual reproduction precludes paedogenesis—an aphid that must mate cannot begin to mature its embryos before it is born (cf. Figure 4.12). This means that a sexually reproducing female must either lay eggs or delay parturition until its offspring can mature. In autumn in temperate regions further increase in numbers is restricted by shortage of food and low temperatures and most aphids enter a resting stage at this time. The production of overwintering eggs provides an excellent opportunity to reproduce sexually, since an egg need not be in an advanced stage of maturation to survive. Aphids have availed themselves of this opportunity: in all holocyclic species the sexual generation is the last in the season—that which produces the resistant eggs (Ward *et al.*, 1984).

Sex ratios

Fisher (1930) demonstrated that in the absence of inbreeding or competition between offspring of one sex for a mate the optimum investment ratio between male and female offspring is unity. This also applies if aphids are regarded as hermaphrodite and egg and sperm production is limited in the main by the same resources (Maynard Smith, 1978). However, the relatively few sex ratios that have been recorded for aphids indicate a preponderance of sexual females over males. This is particularly so in the host-alternating Aphidinae where there can be a ratio as great as 20:1. As the male and female morphs of a clone in these aphids return separately to the primary hosts they are unlikely to colonize the same host and thus outbreeding is certain. In contrast, in the host-alternating Pemphiginae the winged morph (sexupara) that returns to the primary host carries both male and female embryos. These sexuparae colonize the primary host after leaf fall and each rapidly produces 6 to 8 offspring that are arostrate and cannot feed—two of them are males and the rest females. They remain together in a group, the females moult once then mate and lay an egg, and thus inbreeding is certain especially in years when population density is low. The production of two males rather than one is possibly a safeguard

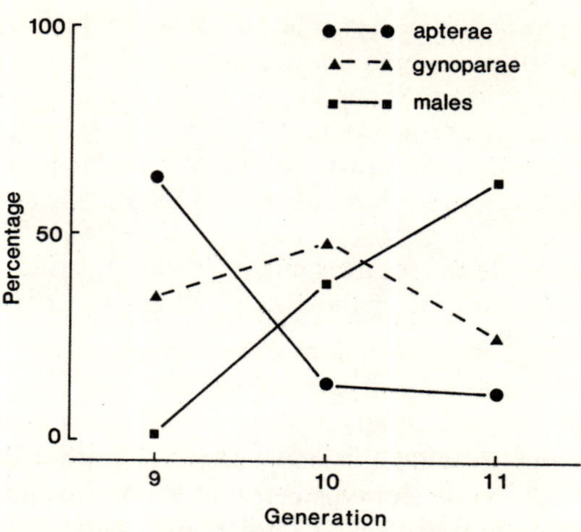

Figure 5.4 The percentage of offspring of *Dysaphis plantaginea* that develop into apterae, gynoparae and males in generations 9 to 11 when reared under autumnal photoperiods. (After Bonnemaison, 1951.)

against accidental loss. In autoecious species the degree of inbreeding is likely to be intermediate between these two extremes and there the sex ratio is usually less than 10:1. Complete and partial inbreeding could account for the distortion of the sex ratio (Hamilton, 1967; Maynard Smith, 1978) in the last two examples. However, the marked excess of females in those species where outbreeding is certain is puzzling.

Several factors could affect the sex ratio in the outbreeding species. In at least one species males are twice as heavy as females at birth and therefore more expensive in resources to produce (Newton, personal communication). Conversely, in terms of time, most host-alternating aphids need twice as long to mature a sexual female as a male. On the initiation of sexual development a gynopara is produced, which then gives rise to oviparae (cf. Figure 6.1). Thus late in the season it is more extravagant in terms of time to produce females than to produce males, and this possibly accounts for the increase in the proportion of males and relative decline in the production of gynoparae towards the end of the season (Figure 5.4). Thus degree of inbreeding and the relative fitness of males and females possibly accounts for the departure from a 1:1 sex ratio and its range in aphids.

Why lay eggs?

Genetic recombination in animals and plants is often associated with a resting stage, with the new genotype(s) emerging from the resting stage at a time and place where they are likely to encounter fresh conditions. Not all aphids overwinter as eggs even in the temperate regions: some overwinter viviparously or as a special morph (hiemalis) that can survive winter without feeding (p. 55). Both eggs and overwintering aphids in temperate regions contain an anti-freeze that may protect them against low temperatures (Sömme, 1969: Parry, 1979*a, b*). However, aphids are not as resistant to low temperatures as eggs.

In south-east Asia and the Mediterranean, areas that do not experience low winter temperatures, the endemic aphids (Greenideinae, Hormaphidinae and Fordini) nevertheless usually overwinter as eggs. In the Mediterranean region the Fordini spend alternate winters in the egg stage and as young parthenogenetic females. Thus if the whole population is synchronized in this way, an egg stage would not appear to be essential for survival.

If, however, plant foliage is largely absent or of very poor quality for a time, and, or, the physical environment is harsh, aphids are in danger of desiccation especially if they live on the above-ground parts of plants. Under these circumstances an egg could be an advantage. In this respect it is interesting to note that the special overwintering morphs produced by a few species overwinter in the soil, where humidity is likely to be high.

Thus the timing of the sexual stage in the life cycle, prior to the onset of harsh conditions, generates a new range of genotypes from which the fittest will be selected when conditions become favourable again. It then becomes a question of how best to survive the adverse conditions. On above-ground parts of plants desiccation may be the most important factor and an egg the best means of combating it. In temperate regions there is the additional advantage that eggs are more resistant than other stages to very low temperatures.

Summarizing, although there is no significant amount of genetic recombination during the development of parthenogenetic eggs, the large size of many aphid populations combined with parthenogenetic reproduction increases the likelihood of a mutation occurring and becoming established. This accounts for some of the adaptability of aphid clones. In addition the sexual reproduction that usually occurs once a year, just prior to the production of eggs, generates genetic diversity that further improves the chances of survival of a species in a heterogeneous and changing environment.

6 Life-history patterns

Aphid life cycles are of two types: autoecious (host-specific) and heteroecious (host-alternating). Autoecious species of aphids live on one or a few species of a particular genus of plants (Figure 6.1); and heteroecious aphids spend autumn, winter and spring on a primary woody host and the summer usually on secondary herbaceous plants (Figure 6.2). The primary and secondary hosts belong to different families of plants (Table 6.1), and the aphids are classified as polyphagous. However, many of these aphids are sequentially monophagous as they live on one plant species at a time, for example, the damson hop aphid alternates between damson and hop.

Members of four of the ten sub-families of the Aphididae (the Aphidinae, Anoeciinae, Hormaphidinae and Pemphiginae) show host alternation (Eastop, 1977). However, even within a genus some species may be heteroecious and others autoecious (Figures 6.1, 6.2), and occasionally the same diversity occurs within a species at the subspecific level (Shaposhnikov, 1981).

Host alternation

The shrubs and trees on which many aphids overwinter as eggs are regarded as the primary hosts because they are phylogenetically older than other hosts of aphids (Mordvilko, 1928). Depending on the genus of aphid, the primary woody hosts can either all belong to the same family or different families of plants. The same is true of the secondary hosts (Table 6.1). Mordvilko (1928) considered heteroecy as an end point in evolution. However, some autoecious species have apparently evolved from heteroecious ancestors. For example, some species of *Cryptomyzus* are heteroecious and migrate from *Ribes*, their primary host, to herbaceous plants of the family Labiatae, their secondary hosts. Other species of this genus live only on Labiatae and are thought to have lost their link with *Ribes* and have become secondarily autoecious (Szelegiewicz, 1978).

As monophagy would appear to have advantages for aphids, there must also be a reason why some species risk leaving their host plants and seek out not only another plant, but a species of another family of plants. Host-alternating species of aphids are often very abundant, despite the risk.

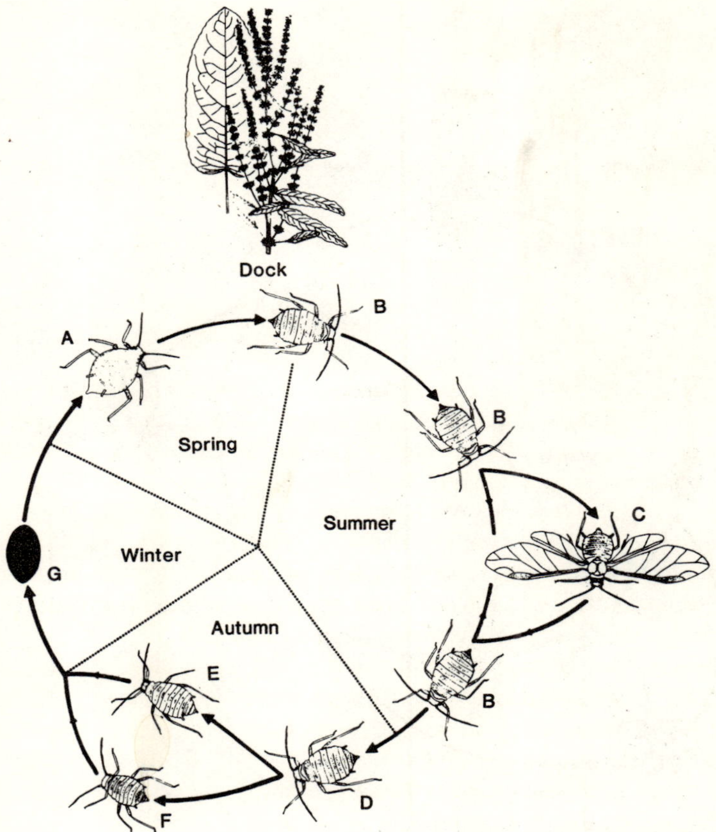

Figure 6.1 Life cycle of the autoecious dock aphid, *Aphis rumicis*. A, fundatrix; B, aptera; C, alate; D, sexupara; E, ovipara; F, male; G, egg. (After Jones, 1942*a*.)

Food quality

While they are growing in spring trees and shrubs are highly nutritious for aphids and again, food quality improves during senescence of the host in autumn. Herbaceous plants tend to grow and flower during summer and supply a favourable habitat when trees and shrubs provide poor nutrition. Both Davidson (1927) and Mordvilko (1928) stressed that, in host-alternating aphids, the seasonal movement between woody primary and herbaceous secondary host plants enables the aphids to exploit a continuous supply of nutritionally favourable foliage that is either growing or senescent.

This idea was extended by Kennedy and Booth (1951), who proposed that host selection by summer generations of the polyphagous black bean aphid (*Aphis fabae*) was in response to nutrient quality, and, that the aphid returned

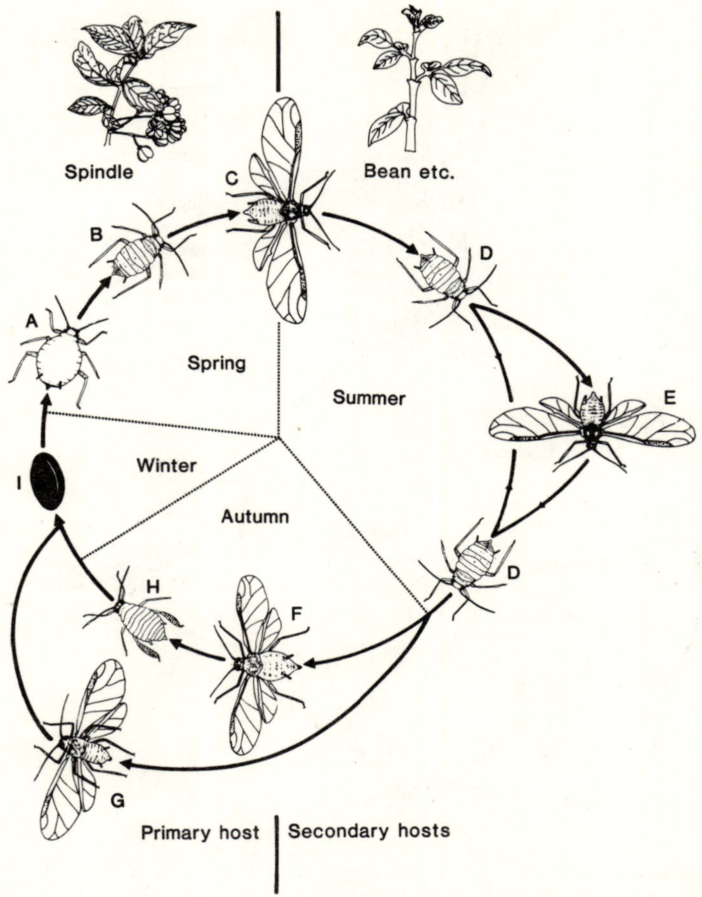

Figure 6.2 Life cycle of the heteroecious black bean aphid, *Aphis fabae*. A, fundatrix; B, fundatrigenia; C, emigrant or spring migrant; D, apterous exule; E, alatae exule or summer migrant; F, gynopara or autumn migrant; G, male; H, ovipara; I, egg.

to the primary host in autumn more in response to secondary plant substances (flavour, p. 11) rather than to nutrient quality. However, this dual discrimination mechanism does not apply to all host-alternating aphids. The preference for primary rather than secondary hosts for a time, and then *vice versa*, in the bird cherry-oat aphid (*Rhopalosiphum padi*) and the apple-grass aphid (*Rhopalosiphum insertum*) is determined by changes in the aphids' flavour preference (Dixon, 1971*c*; Dewar, 1977), as was claimed by Dethier (1947) and Hille Ris Lambers (1945). However, this does not rule out nutrient quality as a major factor in the evolution of host alternation.

There is a big difference in the quality of the food available to aphids feeding on trees and herbaceous plants. Aphids assimilate more energy and achieve

Table 6.1 The plant families of the primary and secondary hosts of three genera of host-alternating aphids (Szelegiewicz, 1978).

	Plant Hosts	
Aphid genus	*Primary*	*Secondary*
Rhopalosiphum	Rosaceae	Gramineae
Pemphigus	Salicaceae	Compositae Gramineae Umbelliferae
Prociphilus	Caprifoliaceae Oleaceae Rusaceae	Pinaceae

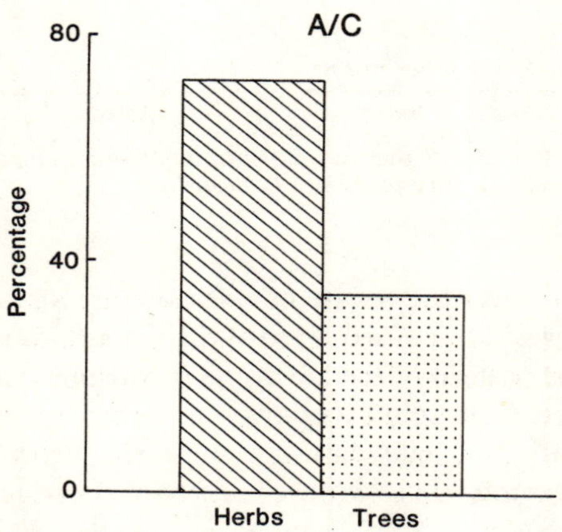

Figure 6.3 The percentage of the energy consumed (C) that is assimilated (A) for aphids that live on herbaceous plants and trees. (After Llewellyn, 1982.)

more growth per unit of sap energy consumed when feeding on herbaceous plants than they do on trees (Figure 6.3) (Dixon, 1975; Llewellyn, 1982), which is possibly a consequence of the lower amino-nitrogen content of the phloem sap of trees (Dixon, 1975). However, the poorer quality of the phloem sap available in trees in summer does not prevent their aphids from achieving high growth and reproductive rates.

Environmental quality

Temperatures can reach high levels in summer, especially in the canopy of trees. In addition to the deterioration in the quality of food, tree-dwelling aphids, more than many other species, experience very high temperatures

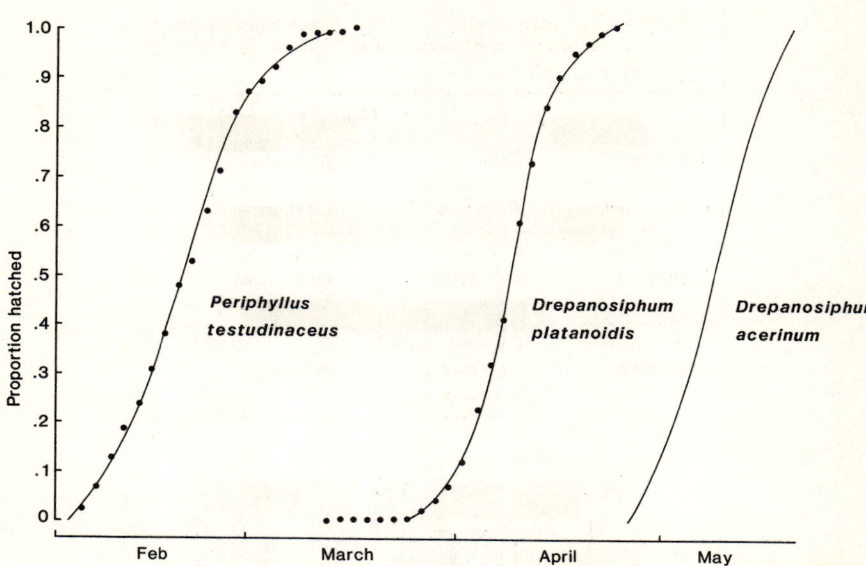

Figure 6.4 The proportion of eggs that have hatched in relation to time of year for *Periphyllus testudinaceus, Drepanosiphum platanoidis* and *D. acerinum.*

during summer and, over the course of a season, are also exposed to a wide range of temperatures. The consequence of this for aphids is possibly revealed by the timing of the pattern of the life cycles of various tree-dwelling aphids.

Of the three species of aphid that live on sycamore, *Periphyllus testudinaceus* hatches early, in late January or early February; followed by *Drepanosiphum platanoidis* in late March and then *D.acerinum* in late April or early May (Figure 6.4). The earliest to hatch, *P.testudinaceus* and *D. platanoidis*, aestivate during summer, the former as a highly specialized first instar nymph (p. 55) and the latter as an adult. The last species to hatch, *D. acerinum*, does not aestivate, but reproduces all through the summer (Figure 6.5). Similar patterns of life cycle occur on other trees. On lime the early-hatching *Patchiella reaumuri* leaves to spend the summer on *Arum maculatum* (Roberti, 1939), whereas *Eucallipterus tiliae* that hatches later, in early May, actively reproduces throughout summer on lime (Figure 6.5). On apple, the aphids *Rhopalosiphum insertum* and *Dysaphis plantaginea* hatch early and then spend the summer on grasses and plantain, respectively, in contrast to the later-hatching *Aphis pomi* that reproduces all through summer on apple and starts producing its sexuals in early autumn, usually before the return of *R.insertum* and *D.plantaginea* (Massee, 1946). On oak *Thelaxes dryophila* hatches early and spends the summer in aestivation, whereas *Tuberculoides annulatus* hatches later and reproduces throughout summer (Figure 6.6).

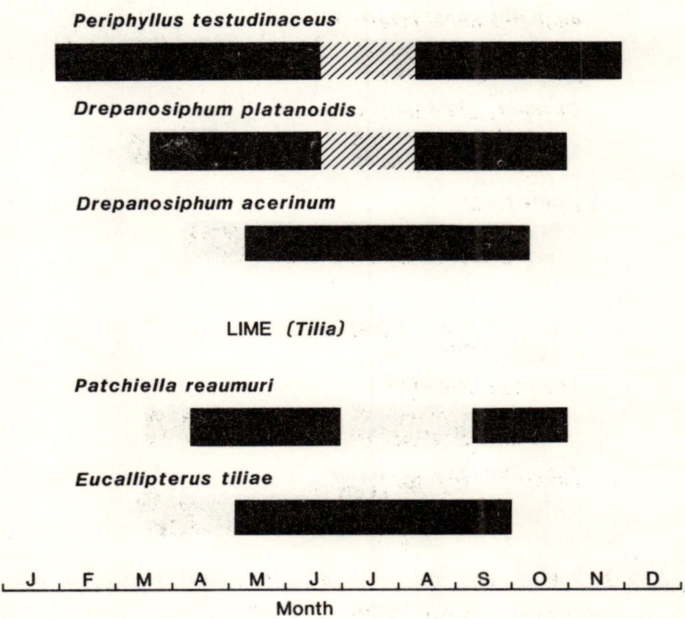

Figure 6.5 The months of the year when aphid species are either growing and reproducing (■), or aestivating (▨) on the tree hosts sycamore and lime, or leave to colonize herbaceous plants (□).

Thus although woody hosts are nutritionally less suitable for aphids during summer (Figure 6.7), nevertheless there are many species of aphids that are still able to grow and reproduce well on trees in summer. This poses the question, why do aphids that overwinter on trees and hatch early either colonize herbaceous plants or aestivate in summer?

Food quality and temperature have a marked effect on aphid size. Poor food quality and high temperatures result in small aphids, as a consequence of the effect of these two factors on aphid developmental and growth rates (p. 23). Young sycamore aphids (*D.platanoidis*) taken from laboratory cultures and caged on sycamore trees in the field in summer suffer a high mortality and only achieve a third of the weight of the smallest individuals recorded in the field. Such small aphids are unable to fly and give birth to very few offspring (cf. Figure 6.10). However, should a leaf senesce in summer then aphids that exploit this high-quality food source grow large and achieve high reproductive rates. Thus in the sycamore aphid aestivation is in response to the harsh conditions of summer. In contrast the later-hatching *D.acerinum* nevertheless continues to develop and reproduce throughout summer.

For each species of aphid there is possibly a limited range of food quality

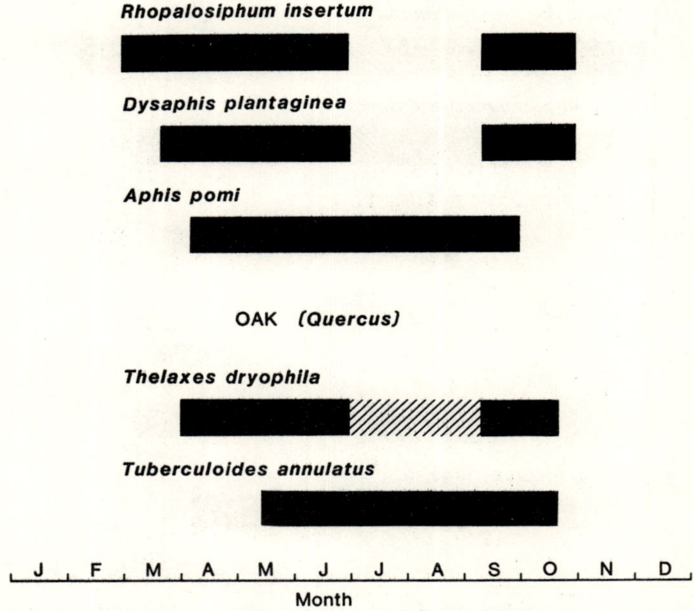

Figure 6.6 The months of the year when aphid species are either growing and reproducing (■) or aestivating (▨) on the tree hosts apple and oak, or leave to colonize herbaceous plants (□).

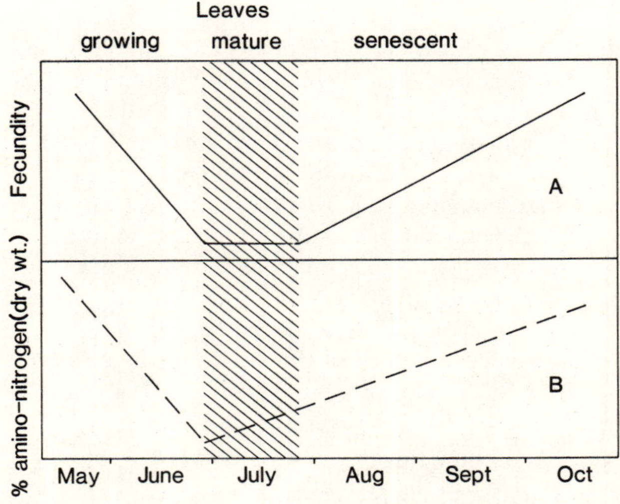

Figure 6.7 The seasonal changes in the fecundity of the sycamore aphid, *Drepanosiphum platanoidis* (*A*) and the amino-nitrogen content of the leaves of sycamore, *Acer pseudoplatanus* (*B*) (After Dixon 1970*a*.)

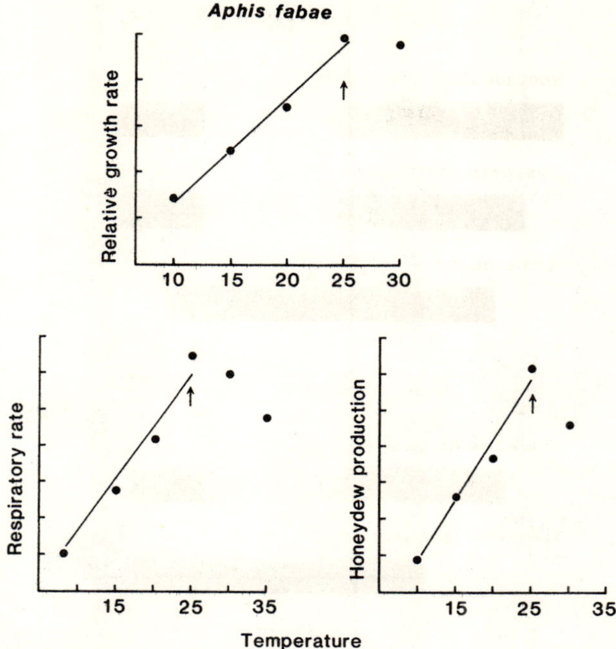

Figure 6.8 The increase in relative growth rate, respiratory rate and honeydew production of *Aphis fabae* in relation to temperature (arrow indicates temperature above which these rates decrease and mortality increases).

and temperature over which it can increase in numbers. At high temperatures an aphid's relative growth rate declines as does its respiratory rate and honeydew production (Figure 6.8), and associated with this is an increase in mortality. However, in addition there also appear to be physiological constraints that prevent an aphid from thriving at both low and high temperatures, with species like *L. erysimi* thriving at higher temperatures than species like *M.euphorbiae* and vice versa (Figure 6.9). The combination of poor nutrition and high temperatures further aggravates the problem. Thus it would appear that tree-dwelling aphids that hatch early are adapted to cool conditions (Dixon, 1973), and are unable to thrive when the host quality is poor and temperatures are high, so they either aestivate or host-alternate. By colonizing herbaceous plants during summer, aphids that hatch early possibly avoid the very high temperatures that occur in tree canopies, especially those aphids that live on the roots of their secondary hosts, and benefit from the higher nutritional quality of herbaceous plants. In this context it is interesting that several species of host-alternating Pemphiginae spend the summer not on herbaceous plants but on the *roots* of other species of trees, mainly of Gymnosperms. Several species of the autoecious Lachninae leave the canopy

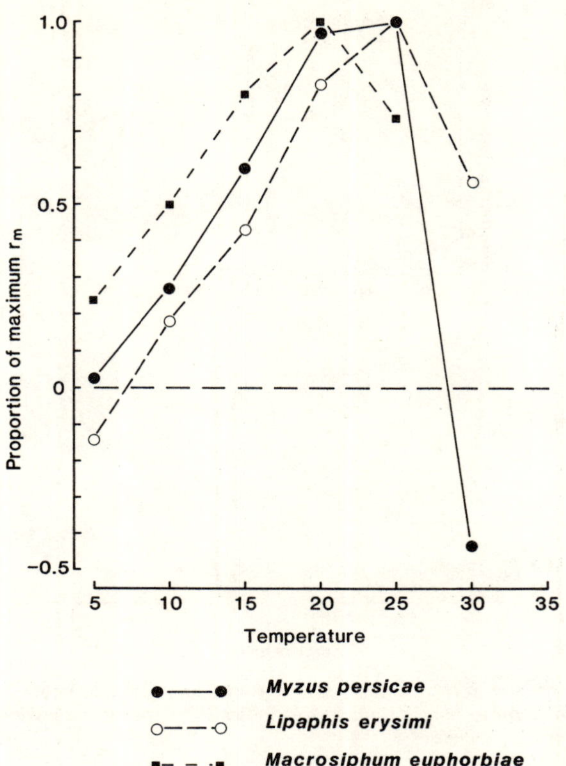

Figure 6.9 The proportion of the maximum natural intrinsic rate of increase achieved in relation to temperature for three aphids. *Macrosiphum euphorbiae* does relatively better at low than high temperatures compared with *Lipaphis erysimi* (After Barlow, 1962; De Loach, 1970.)

and live on the *roots* of their coniferous host trees during summer. Therefore, as is well illustrated by the sycamore aphid, although canopy temperatures during summer do not reach lethal levels, nevertheless the aphid is not adapted to grow and reproduce on the poor-quality nutrition and at the high temperatures that prevail in summer (Figure 6.10). This is surprising as the summer morphs of several species show differences in, for example, the length of their gut or proboscis (Figure 6.11) and fat content that enable them to survive the conditions they are most likely to experience (p. 79).

Resource partitioning

This temporal patterning in aphid life cycles might also be explained in terms of resource partitioning followed by morphological and physiological adaptations to the particular conditions. However, there is no evidence for this. As *D.acerinum* is not present throughout most of the range of *D.platanoidis*, it is

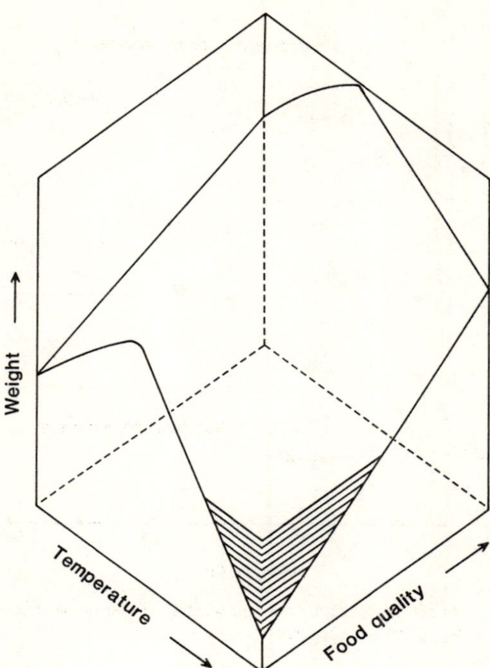

Figure 6.10 The weight of the sycamore aphid, *Drepanosiphum platanoidis*, in relation to temperature and food quality. High temperatures and poor food result in the development of very small adults (▨).

difficult to understand why the latter has not exploited the summer months, unless it is constrained by its physiology.

Natural enemies

An alternative hypothesis, that host-alternation is a means by which some aphids escape their natural enemies, has attracted little attention. When a large aphid population has developed on a primary host, some of the winged forms (emigrants, Figure 6.2) that develop escape the accumulation of natural enemies by moving to other, secondary, host plants. The new colonies, because of their isolation from other colonies, are unlikely to be located quickly by enemies. While the aphid remains undetected it can rapidly increase in numbers on the rich food supply and produce more winged aphids that in turn colonize other plants. Movement from plant to plant would enable the aphid to exploit the spatial heterogeneity that exists in nature, and so escape its natural enemies for the time being (Dixon, 1971*c*). The absence of *A.fabae* from spindle during the summer has been attributed to the activity of natural enemies (Way and Banks, 1968). Predator and parasite activity

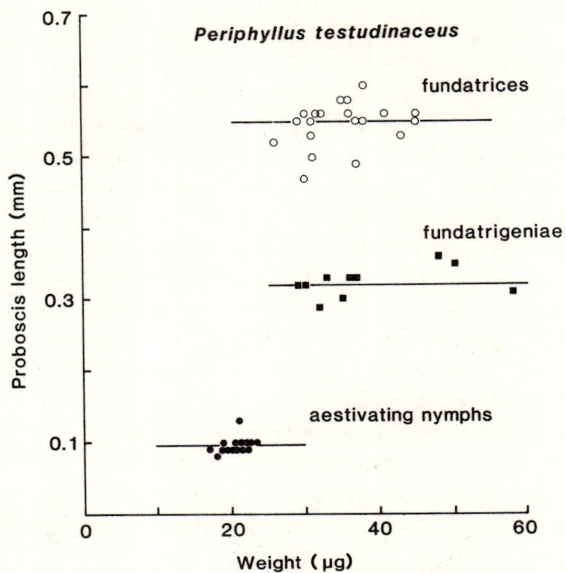

Figure 6.11 Proboscis length in relation to total weight of the fundatrices, fundatrigeniae and aestivating nymphs of *Periphyllus testudinaceus*.

however, cannot account for either the relationship between the time of egg-hatch and the occurrence of host-alternation, or, the relative rarity of host-alternation. Thus the activity of natural enemies cannot have been the major factor favouring the evolution of host alternation. However, natural enemy activity may well have partly determined the timing of the migrations between primary and secondary hosts, especially of the more vulnerable gregarious species.

Although trees and shrubs provide a harsh environment for aphids during summer some species are nevertheless adapted to thrive under these conditions. Host-alternation, like aestivation, is a means of bypassing these harsh conditions. The relative rarity of host-alternation may be a consequence of the difficulty in adapting to feed and live on two or more, as opposed to one, species of host plant, and the availability of other common species of plants to exploit as hosts. The seasonal changes in plant quality and temperature, so characteristic of temperate regions, and the physiological constraints on aphid development, possibly account for the diversity of life-history patterns. Little is known about the biology of aphids living in the tropics but it is tempting to suggest that the poor aphid fauna there is partly due to the fewer niches that exist there for aphids because of the constancy of the tropical environment.

Seasonality

Aphids of each of the generations that make up a life cycle have specific reproductive 'strategies' adapted to the conditions they are most likely to encounter: in effect they anticipate the predictable seasonal trends in habitat quality. This can be illustrated by reference to the sycamore aphid, an autoecious species, and the black bean aphid that is heteroecious.

The sycamore aphid has only three morphs: alate viviparous females, apterous oviparous females and alate males. The second-generation adults, that develop and are present in summer, enter a reproductive diapause and have a low fecundity thought to be a direct consequence of the poor quality of the food available to the aphid when feeding on mature leaves (Mordvilko, 1908). However, when reared at low temperatures on unfurling leaves, conditions normally experienced by the first generation, the large, second-generation individuals that develop have well developed fat bodies and poorly developed gonads even so, as if about to enter aestivation.That is, they are prepared for the onset of harsh conditions associated with the cessation of growth of the leaves of sycamore (Dixon, 1975*b*). Second-generation aphids also differ from those of the first generation in having fewer ovarioles (Wellings *et al.*, 1980), larger appendages and a greater number of rhinaria on their antennae (Dixon, 1974*b*), a longer gut (Dixon, 1975*b*) and a higher wing-beat frequency, and they fly longer and lower (Mercer, 1979).

In autumn when temperatures are low and nutrients are being recovered from the leaves the quality of the habitat of the sycamore aphid is also good. Thus the quality of the habitat in which the spring (the first-) and autumn (the third-) generation individuals live is good, whereas summer (second-) generation individuals live in a relatively poor habitat.

Let us summarize some generation-specific characteristics. First- and third-generation individuals have a good habitat, and associated with this, a short adult life, small fat body, weak spacing behaviour and produce many small offspring quickly—features characteristic of an *r*-selected species. Second generation individuals have a harsh habitat, a long adult life, well-developed fat body, marked spacing behaviour and produce fewer and larger offspring—features characteristic of a *K*-selected species. The development of these generation-specific characteristics is not affected by rearing the aphid under constant conditions (Dixon, 1975*b*; Leather and Wellings, 1981).

Life cycles of host-alternating aphids like the black bean aphid are more complicated. *A. fabae* overwinters as an egg on spindle (*Euonymus europaeus* L.), hatching in spring and giving rise to two generations of apterous individuals, followed by a third generation of alatae. The alatae colonize broad beans and other herbaceous plants. The alate exules produced on the secondary hosts then colonize other secondary hosts, and the males and gynoparae that are produced in autumn migrate to spindle. The gynoparae give birth to apterous oviparae that lay the overwintering eggs after mating with the males (Figure 6.2).

The trend in the reproductive traits of the morphs of this aphid tend to be similar on both the primary and secondary hosts. Spindle at bud burst supplies the first generation aphids with a rich food supply that becomes less suitable as spindle ceases growing. First-generation aphids have large gonads and rapidly produce a large number of small offspring; whereas second-generation aphids have smaller gonads and produce fewer, but larger, offspring. Although small at birth, the offspring born to first-generation adults achieve a high growth rate on the rich food supply available to them and become large adults. The emigrants migrate from spindle to actively growing beans which are a very nutritious host compared to spindle by this time. However, beans soon cease growing and lose quality. Therefore, on both the primary and secondary hosts it is an advantage for the later generations to produce larger offspring, as they are more likely to become adults fit to survive the harsh nutritional conditions, than are small offspring. On both primary and secondary hosts, individuals in the later generations also tend to have more fat. Therefore, as in the sycamore aphid, individuals of the first generation put proportionally more of their resources into reproduction, and less into fat, than individuals of later generations do.

The alate exules that disperse and colonize fresh secondary hosts late in summer also initiate a sequence of morphs whose reproductive traits are similar to those of the sequence of morphs initiated by emigrants. However, it is not done by varying the size of the gonads, which all have 12 ovarioles. The offspring of alate exules grow faster, have more well developed embryos in their gonads and a higher reproductive rate than aphids of subsequent generations. These characteristics are also adaptive since alate exules colonize plants that are nutritionally more favourable than the ones on which they developed (p. 89).

Attempts to change these characteristics by rearing aphids in constant conditions and on high-quality host plants have resulted in a reversal of the downward trend in size from generation to generation, but no change in the generation-specific reproductive characteristics (Figure 4 in Dixon, 1980*b*; Leather and Wellings, 1981).

Generation-specific 'strategies'

The *r*-selected characteristics of aphids, in particular their high intrinsic rate of increase, have been appreciated. It has been argued that in aphids that show alary dimorphism the apterae are more 'juvenile' and have a marked vegetative function, whereas the alatae are more 'adult' and have a marked locomotory function (Kennedy and Stroyan, 1959). These morphs have been classed by Ito (1980) as *r*-and *K*-selected, respectively.

The generation-specific plasticity in the expression of a genotype may be as important to the success of aphids as the opportunistic multiplicative faculty of parthenogenesis so often cited. The generation-specific strategies evolved in the sycamore and black bean aphids enable them to anticipate seasonal changes in habitat quality, while extrinsic variables tune, through phenotypic plasticity, the reproductive strategy to the actual conditions experienced (Wellings *et al.*, 1980; Dixon and Wellings, 1982).

Currently life-history strategies are classifed as species characteristics (Stearns, 1976), and individual differences in life-history strategies within a species are explained on a genetic basis. At present, however, a single genotype, with seasonally expressed adaptations, has no place within the general framework of life-history theory, centred as it is on the ideas of *r*-and *K*-selection.

Life-history theory is largely dependent on comparative studies of closely related species living in different habitats (Bell, 1980). Each species has a set of co-adapted characters appropriate for living in a particular habitat. It has proved difficult to dissociate these characters experimentally and thus determine their relative costs and benefits for a species. However, there are many aphid species that are ideal for testing life-history theory, especially those that show intrinsically programmed and variable reproductive strategies within a morph.

Habitat predictability and tactical diversity

At least four alternative strategies have evolved in animals to maximize fitness: pessimism, phenotypic plasticity, spreading the risk in space and time, and mixed strategies (the production of phenotypically diverse offspring). Examples of all but the first strategy have been described for aphids.

Let us assume that aphids are food-limited. So far no one has been able to demonstrate that food is limited because it is difficult to measure the quality and quantity of food available to aphids. However, the inverse relationship between mortality and relative growth rate recorded for several species (Figure 6.12) supports this contention, as the highest growth rates are rarely achieved. In addition the large sycamore aphids that develop when reared at 10°C show signs of nutritional stress when transferred to 20°C, compared with the controls (Figure 6.13). Their disproportionately large embryonic mass requires more nutrients than are available to the aphids, and this affects both survival and rate of increase, and the birth weight of their offspring (Wellings and Dixon, 1983).

Theoretically, as there are nutrition-related costs in terms of increased mortality, and benefits, in increased reproduction, if there are more ovarioles in an aphid's reproductive system (Figure 6.14), then seasonal trends in host quality should be reflected in a trend in ovariole number. Fundatrices, present when plants are most nutritious, generally have more ovarioles than individuals of following generations. For example, fundatrices of *Thecabius*

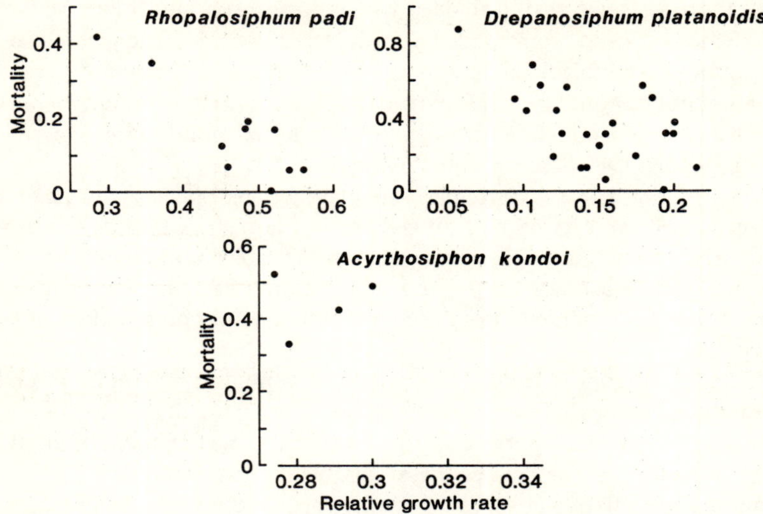

Figure 6.12 Mortality in relation to relative growth rate for *Rhopalosiphum padi, Drepanosiphum platanoidis* and *Acyrthosiphon kondoi*. (Wellings, pers. comm.)

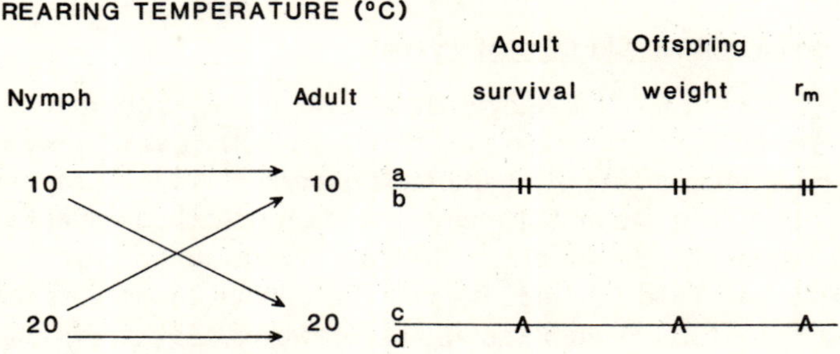

Figure 6.13 The consequence of rearing the sycamore aphid, *Drepanosiphum platanoidis*, at either 10 or 20° C and on becoming adult of transferring them to the other temperature (*b, c*) or keeping them at the same temperature (*a, d*). Aphids transferred from 10 to 20° C have a poorer adult survival, produce smaller offspring and have a lower rate of increase than the control aphids (After Wellings and Dixon, 1983.)

affinis of the subfamily Pemphiginae have 50 ovarioles, in contrast to the 13 recorded for later generations (Wellings *et al.*, 1980; Tashev and Markova, 1982). There are few exceptions: these include the aestivating adults of the sycamore aphid that live for a very long time and span periods of very poor and very rich nutrition but nevertheless develop a relatively large number of ovarioles. During the period of nutrient stress, the reproductive biomass, that is to say the mass of embryonic tissue, which determines the stress-related mortality, is very low. Afterwards, when the host quality has shown a

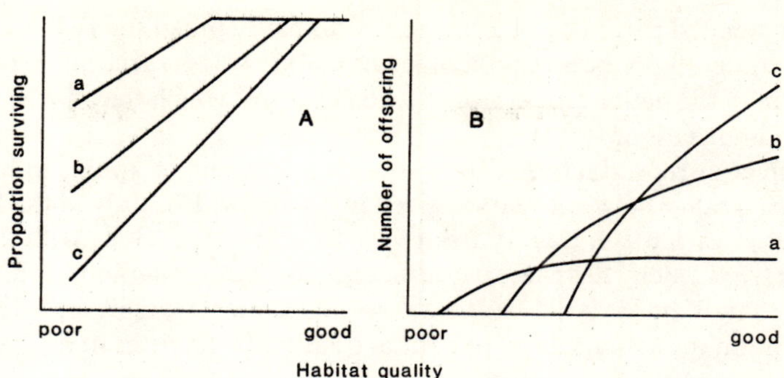

Figure 6.14 The proportion of aphids surviving (*A*) and the number of offspring they have (*B*) in relation to habitat quality and a small (*a*), medium (*b*) and large (*c*) reproductive investment. (After Ward *et al.*, 1983*a*, *b*.)

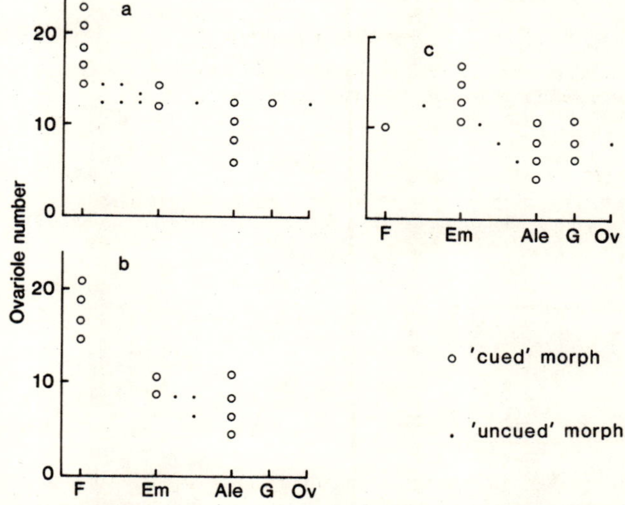

Figure 6.15 The range of ovariole numbers in 'cued' and 'uncued' generations of three species of aphid. (*a*), *Aphis fabae*; (*b*), *Metopolophium dirhodum*; (*c*) *Rhopalosiphum padi*. (After Dixon and Dharma, 1980*b*; Wellings *et al.*, 1980.)

dramatic improvement, the same adults are able to capitalize on the very rich nutrition by virtue of their large number of ovarioles. Thus, in the few species where an individual lives long enough, selection can apparently act separately on reproductive biomass and ovariole number.

As the quality of host plants fluctuates seasonally, any variation in the time of appearance of an aphid morph will mean that the morph will encounter variation in host quality from year to year. During the year, aphid morphs whose production is determined ('cued') by environmental stimuli will be better synchronized than other morphs. Generations or morphs that appear over an extended period can be considered to have spread the risk over time. The risk trade-off predicts that these forms will show less variation in ovariole number than the better synchronized, cued morphs, which they do, at least in the Aphidinae (Figure 6.15).

The effect of reproductive strategy on overall fitness of aphids in habitats varying in quality between years is given in Table 6.2. It is only in the habitat that is most variable in quality from year to year that a clone with a mixed reproductive strategy has the greatest fitness. In all other habitats aphids with either a small or a large reproductive investment would be favoured, dependent on the average level of habitat quality. Differences in reproductive

Table 6.2 The effect of reproductive strategy on overall fitness of aphids in habitats varying in quality between years

(a) *Effect of reproductive investment and habitat quality on fitness within a year*

Reproductive investment (strategy	Habitat quality	
	poor (P)	good (G)
High	0	4
Low	1	1
Mixed	0.5	2.5

(b) *Effect of change in habitat quality from year to year on overall fitness*

Reproductive investment	Range in habitat quality within and between years			Overall fitness
	G G / G G	⟷	P P / P G	
High	$\bar{4}$ } fitness	×	$\bar{1}$	4*
Low	$\bar{1}$ } fitness	×	$\bar{1}$	1
Mixed	2.5 } fitness	×	$\bar{1}$	2.5
	G G / G P	⟷	P P / P G	
High	$\bar{3}$	×	$\bar{1}$	3*
Low	$\bar{1}$	×	$\bar{1}$	1
Mixed	$\bar{2}$	×	$\bar{1}$	2
	P P / P P	⟷	P P / P G	
High	$\bar{0}$	×	$\bar{1}$	0
Low	$\bar{1}$	×	$\bar{1}$	1*
Mixed	$\overline{0.5}$	×	$\bar{1}$	0.5
	G G / G G	⟷	P P / P P	
High	$\bar{4}$	×	$\bar{0}$	0
Low	$\bar{1}$	×	$\bar{1}$	1
Mixed	$\overline{2.5}$	×	$\overline{0.5}$	1.25*

*most fit.

attributes between species, especially from different subfamilies, however, may be due to differences in selection pressure or to differences in the constraints on their ability to produce phenotypically diverse offspring. That the evolution of mixed reproductive strategies has not been found in the Drepanosiphinae argues strongly for a phylogenetic constraint (Ward and Dixon, 1984).

Response to habitat quality

Some aphids can respond to changes in habitat quality by adjusting the proportion of their offspring that is committed to a high or a low level of reproductive effort. The greater the nutrient stress experienced by mothers, the fewer of their offspring are committed to a high reproductive investment, and vice versa (Figure 6.16) (Walters and Dixon, 1985). Even in a poor-quality habitat, however, some offspring will have a large number of ovarioles, and vice versa (Figure 6.17). Therefore the degree of association between weight, an indicator of habitat quality, and reproductive investment will rarely be good.

Thus, to some extent, certain aphids can tailor the reproductive investment of their offspring to fit trends in habitat quality more closely, maintaining a high reproductive rate in rich habitats and reducing mortality in the poorer habitats. The retention of a range of ovariole numbers in all habitats possibly reflects an advantage of spreading the risk between individuals of a clone, with some allocating more of their resources to dispersal and survival than reproduction, and vice versa.

Some aphids show flexibility of reproductive capability, in that nutrient stress promotes embryo resorption. Because of its short life an individual aphid on a poor host cannot usually expect any great improvement in habitat quality and it makes the best use of what is currently available: it matures its

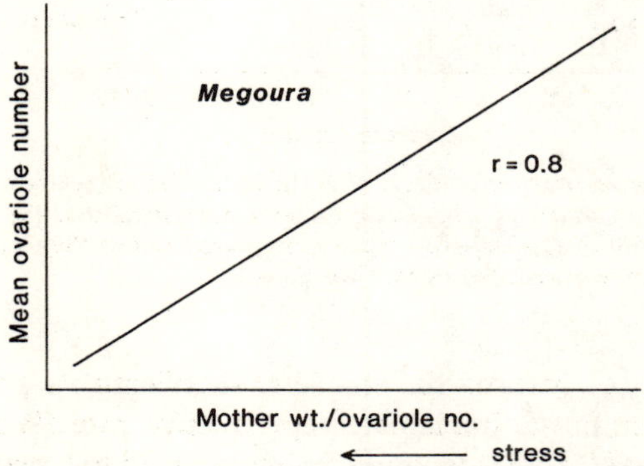

Figure 6.16 Mean number of ovarioles (reproductive investment) of offspring in relation to the weight divided by the ovariole number of the parent in *Megoura viciae*. Low values for the weight divided by number of ovarioles indicates a high degree of stress and *vice versa*.

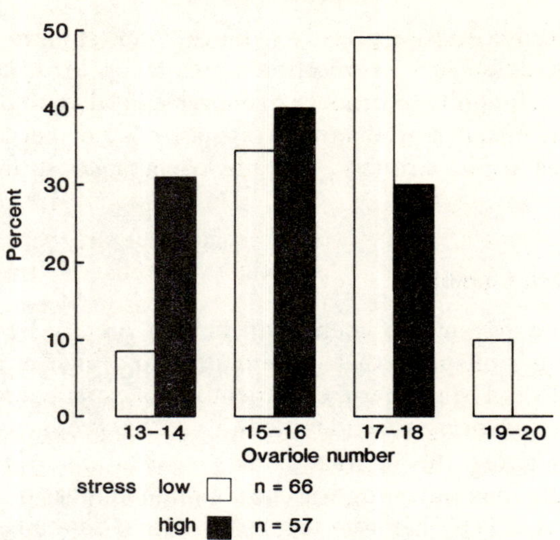

Figure 6.17 Percentage of offspring with 13–14, 15–16, 17–18 and 19–20 ovarioles in relation to the nutritional stress (cf. Figure 6.16) experienced by the parents.

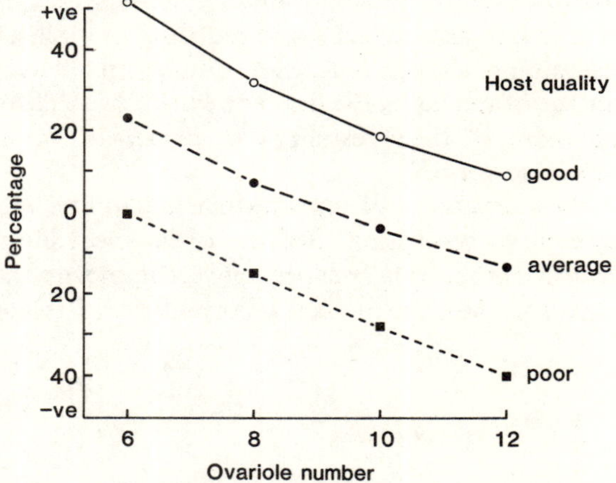

Figure 6.18 Percentage of embryos ovulated after the adult moult in relation to ovariole number and the quality of the habitat in *Aphis fabae*. In poor habitats all but the aphids with lowest reproductive investment resorb embryos, whereas in good habitats they all ovulate additional embryos but this is inversely related to ovariole number.

largest embryos and resorbs the smallest, sacrificing its potential lifetime fecundity to maintain its immediate reproductive rate (Ward and Dixon, 1982; Leather *et al.*, 1983). In good habitats an aphid may ovulate extra embryos (Figure 6.18) and so increase potential fecundity (Ward *et al.*, 1983).

Thus to a very limited extent, aphids are indirectly able to adjust their reproductive biomass and escape the constraints imposed by decisions made during early development.

Summarizing, aphids that live on shrubs and trees during summer experience a poor food supply and high temperatures. For the aphid species that overwinter on trees and shrubs and hatch early in the year, host alternation and aestivation are means of escaping the harsh summer conditions. However, the species that hatch later are better adapted to summer conditions and are able to thrive on shrubs and trees despite the environmental changes. Aphids 'anticipate' the seasonal changes in habitat quality by producing a sequence of morphs that are each adapted to the conditions they are likely to experience. In addition, in some subfamilies aphids show intra-clonal tactical diversity in reproductive investment, which possibly helps them to survive the year to year heterogeneity in habitat quality. By responding also to current conditions aphids can by means of maternal control adjust the reproductive commitment of their offspring, and by ovulation or resorption of embryos during adult life, alter their reproductive capability and so modify decisions made earlier in their development.

7 Dispersal

Most species of aphids, at all stages of development, move about over the surface of their host plants and even between adjacent plants. These local movements result in slow diffusive dispersal. In contrast, aphids also show persistent 'straightened out' movements during which their vegetative responses are depressed, that are a means of transport over greater distances (Kennedy, 1975). This latter behaviour is characteristic of the winged stages, but is also shown by nymphs. This is well illustrated by *Pemphigus trehernei* that lives on the roots of sea aster that grows on mud banks in tidal salt marshes. Unfed first instar nymphs become active and photopositive. They leave the roots and climb up to wander on the exposed soil surface. The rising tide lifts and drifts the nymphs which, on standing erect on the water surface, are caught by the wind which sends them scudding across open water to be deposited by the falling tide on another mud bank. Thirty minutes of this waterborne experience is enough to reverse the reaction of these nymphs to light and send them down a new crevice where aster roots may be found and colonized. This reversal to photo-negative behaviour provides some indirect control over the distance the nymphs travel (Foster, 1978; Foster and Treherne, 1978). Thus local, 'trivial', movements and distant, 'migratory', movements can be recognized, and it is possible that they are the extremes of a continuum of movement that disperses aphids.

Adaptive significance

In addition to the assumed high mortality suffered during dispersal, alate aphids incur other disadvantages that are more easily quantified. The development of a flight apparatus and/or its maintenance is costly in terms of time and fecundity (p. 47). If aphids fly then they may incur an additional cost in that their potential fecundity is further reduced (Burns, 1971) and there is a further delay in the onset of reproduction. The combined effect is a marked reduction in reproductive potential and rate of increase.

Although dispersal results in the colonization of suitable plants (Dixon, 1969) it is not always clear what advantages there are in dispersing except from

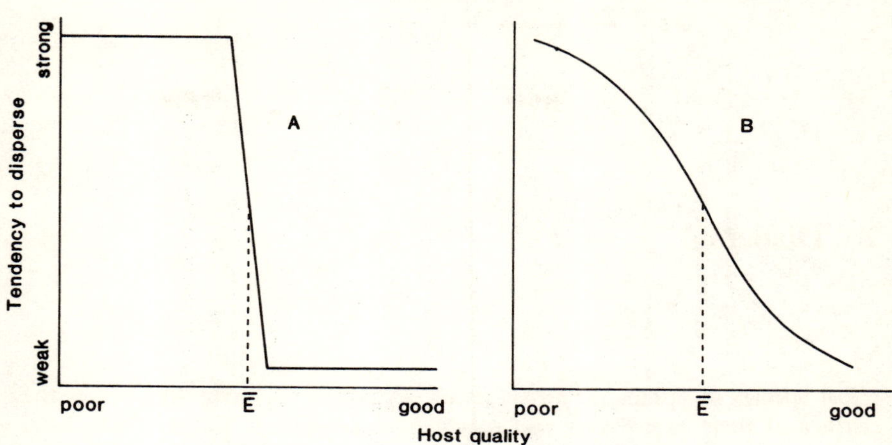

Figure 7.1 Tendency to disperse in relation to habitat quality ($\bar{E}$, average expectation).

annual plants that are about to die. As plants continually vary in quality, aphids, however, could profitably colonize a new host plant when the quality of the plant on which they are feeding falls below the average for the habitat (Figure 7.1). This applies whether they live on more permanent hosts like trees or on temporary hosts like herbaceous plants (Dixon, 1969). As dispersal can be costly in terms of fecundity or survival, or both, then aphids are likely to delay departure until host quality falls well below the average expectation for the habitat. Aphids that do not respond in this way are likely to leave fewer offspring. However, changes in abundance, distribution and quality of host plants from year to year alter the average expectation. This raises interesting questions about the best way to cope with environmental heterogeneity (p. 81).

However it is, in fact, adaptive to produce some winged individuals that leave when host quality is still above the average expectation, because it serves to increase the chance of survival by dispersing members of a clone in space. Thus there is rarely a dramatic switch as illustrated in Figure 7.1*A*, but a more gradual change, with migratory and non-migratory individuals present over a wide range of host quality (Figure 7.1*B*).

Take-off

Accidental dispersal is probably relatively rare in aphids. Polymorphic species have to change from apterous to alate development, and this ontogenetic switch takes time and is usually in response to overcrowding or a deterioration in host quality, or both (p. 38). Once alates have settled on a host plant they autolyse their wing muscles (Figure 7.2), which further reduces any chance of accidental dispersal.

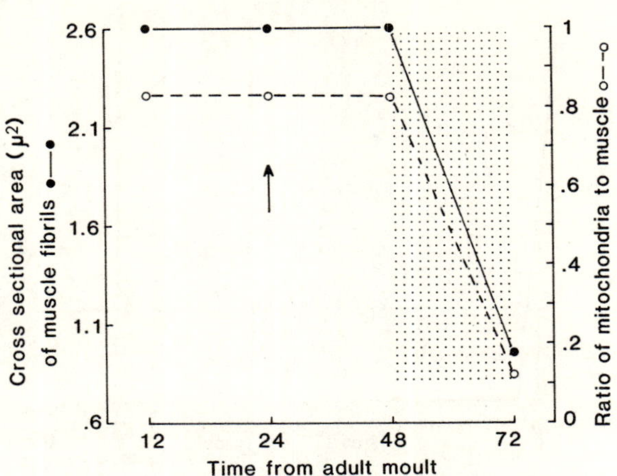

Figure 7.2 Changes in the cross-sectional area of the muscle fibrils and the ratio of mitochondria to muscle area in transverse sections through the dorso-ventral indirect flight muscles of *Megoura viciae* in relation to the time from adult moult (↑ flight usually occurs at this time, ⬚ wing muscle autolysis). (After Burns, 1971.)

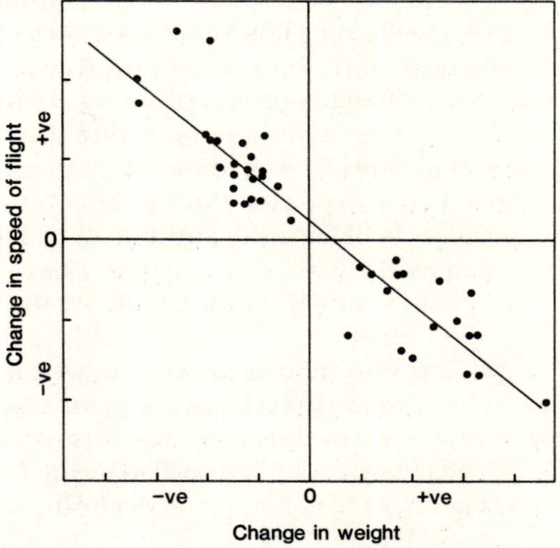

Figure 7.3 Changes in flight speed in the sycamore aphid, *Drepanosiphum platanoidis*, in relation to changes in weight achieved by keeping the aphid on good- and poor-quality hosts. (After Mercer, 1979.)

The viviparae of some tree-dwelling aphids are all winged and can fly throughout adult life, but even in these species accidental dispersal is unlikely. When feeding on a good-quality host, sycamore aphids increase in weight and vice versa (Figure 7.3) and can only fly from a good-quality host after reducing their weight. This is achieved by excreting several drops of honeydew.

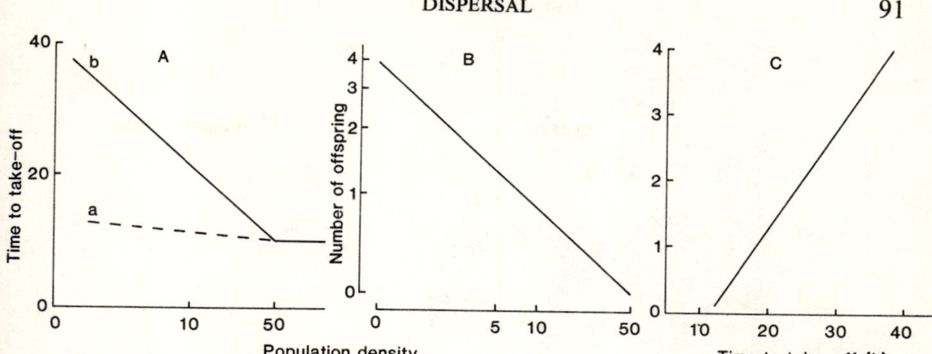

Figure 7.4 Time to take-off in hours (*A*) and the number of offspring produced before flying (*B* + *C*) by *Sitobion avenae* in relation to population density (*A* + *B*) and time to take-off (*C*). *a* = poor-quality host, *b* = high-quality host. (After Walters and Dixon, 1984).

Crowding markedly influences the readiness with which aphids fly (Figure 7.4*A*). At low levels of crowding, take-off is delayed and offspring are produced before an aphid flies (Figure 7.4*B*). On a good host some species of aphid will not take off unless crowded, whereas in others a crowding stimulus is not required. Host quality also affects the readiness to take off, and aphids generally depart sooner from poor-quality hosts (Figure 7.4*A*), and the shorter the stay the fewer offspring they are likely to leave behind (Figure 7.4*C*).

Thus, although capable of flight, alate aphids respond to current cues of host quality. They leave quickly if it is poor, but if it remains above the average expectation they either deposit young before flying, or stay. This facultative dispersal is characteristic of aphids living on plants that vary in quality during the season. However, when the trend in quality is predictable obligate migrants are produced. The emigrants and sexuals of host-alternating aphids (p. 70) are produced in anticipation of an inevitable deterioration in host quality. These morphs are obligate migrants (Dixon, 1971*b*), and colonize other more suitable hosts.

Physical factors also affect take-off. Aphids are more likely to leave at high than low light intensities and temperatures, and at low rather than high wind speeds (Figures 7.5, 7.6). Species differ in their response to these factors. The sycamore aphid flies at lower light intensities and temperatures than the lime aphid. This partly determines their different daily flight activity curves (cf. Figure 7.11). Take-off in tree-dwelling aphids, such as the sycamore aphid, is more markedly inhibited by high wind speeds than is the take-off of aphids that live on short-lived host plants (Haine, 1955; Dixon and Mercer, 1983; Walters and Dixon, 1984). Delay is more critical when living on short-lived hosts that are likely to deteriorate rapidly in quality. Even in these species high winds may inhibit take-off for a day, but not longer, indicating a change in threshold. However, although it is not known whether other tree-dwelling

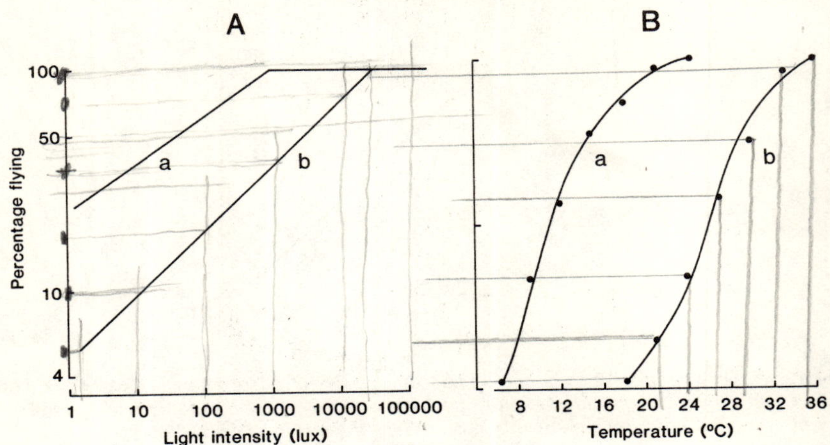

Figure 7.5 Percentage of sycamore (*a*) and lime aphids (*b*) flying in relation to light intensity (*A*) and temperature (*B*).

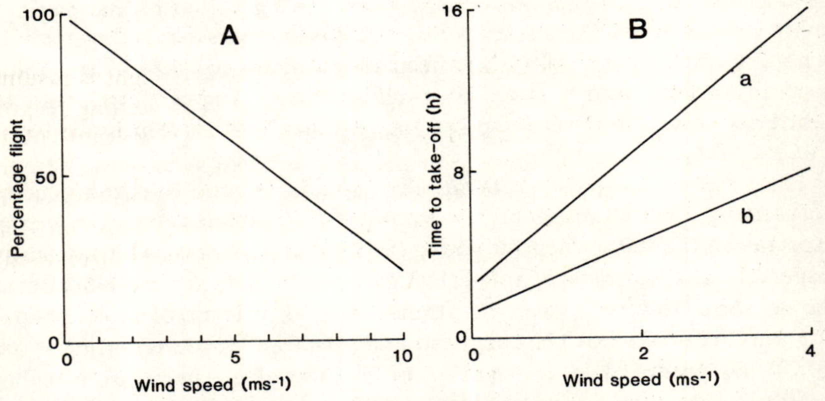

Figure 7.6 Percentage of sycamore aphids flying in 24 hours (*A*) and time to take-off of the cereal aphid, *Sitobion avenae* (*B*), from good- (*a*) and poor-quality hosts (*b*) in relation to wind speed. (After Mercer, 1979; Walters and Dixon, 1984.)

aphids share the sycamore aphid's reluctance to fly on windy days it is possible that this behaviour is associated with tree dwelling, that is, living on a relatively permanent host. In general then, adverse weather conditions are only likely to delay rather than prevent take-off (Johnson, 1969; Walters and Dixon, 1984).

Settling

Aphids are more likely to settle on a good- than on a poor-quality host (Walters and Dixon, 1982). If an aphid settles on a non-host plant or poor-

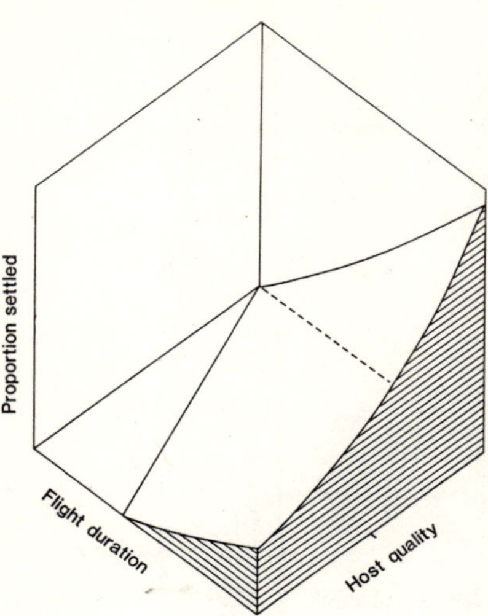

Figure 7.7 Proportion of aphids settled in relation to flight duration and host quality. (After Johnson, 1958; Kennedy and Booth, 1963*a*.)

quality host the settling response is quickly inhibited and flight is resumed. Repeated flights have a cumulative positive effect on the settling response (Figure 7.7) (Johnson, 1958; Kennedy and Booth, 1963*b*). The length of time for which an aphid has flown, and the number of plants it has sampled, possibly give some indication of the abundance of potential hosts. The optimal strategy after a long and unsuccessful search would be to settle and feed on the next host plant encountered because of the low likelihood of finding a better host.

Tendency to disperse

At the population level, the tendency to disperse has been measured in terms of the proportion of a population that develops into alatae (Lamb and MacKay, 1979), and at the individual level, in terms of whether an aphid flies before or after giving birth to offspring (Shaw, 1970*b*). Crowding during nymphal life affects the development of wings, flight musculature, fuel reserves and gonads. Consequently, migrants have a well-developed flight apparatus but poorly-developed gonads—the oogenesis flight syndrome (Johnson, 1969; Shaw, 1970*c*). Flight is partly dependent on crowding experience during nymphal (Shaw 1970*b, c*) and early adult life (Dixon *et al.*, 1968; Walters and Dixon, 1982).

However, variability in the tendency of individuals to disperse is also associated with programmed differences in the level of reproductive investment (number of ovarioles in their gonads, p. 46) between individuals within a clone, and this is independent of their size (p. 80), those with few ovarioles

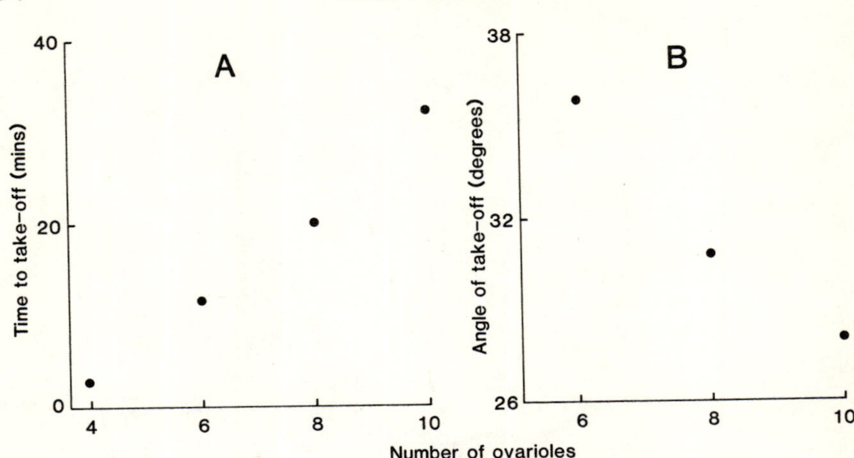

Figure 7.8 Time to take-off (*A*) and angle of take-off (*B*) in relation to the number of ovarioles (reproductive investment) in *Sitobion avenae*. (After Walters and Dixon, 1983.)

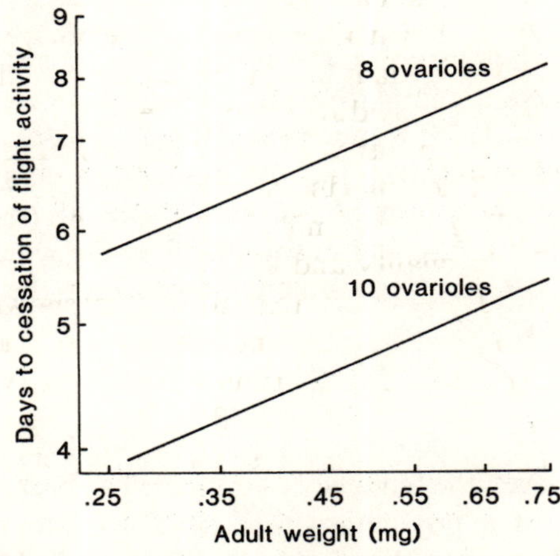

Figure 7.9 Days to the cessation of flight activity (wing muscle autolysis) in relation to adult weight in *Sitobion avenae* with eight and ten ovarioles. (After Walters and Dixon, 1983.)

take off more readily and at a steeper angle (Figure 7.8), delay wing muscle autolysis longer (Figure 7.9), are more resistant to starvation, and have relatively more olfactory organs than aphids with many ovarioles (Walters and Dixon, 1983). In *A. fabae* the proportions flying for increasing lengths of time (Kennedy and Booth, 1963*a*) are similar to the proportions that have decreasing numbers of ovarioles (Figure 7.10). Similarly, the proportions of alates of *A. fabae* that take off before giving birth, after giving birth or that do not fly at all (Shaw, 1970*b*) may also be associated with reproductive

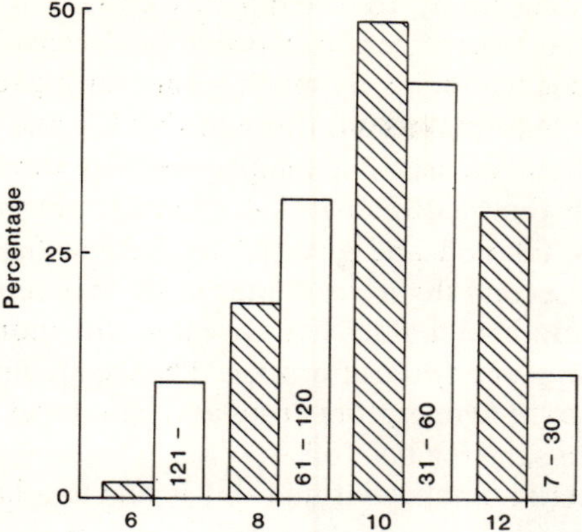

Figure 7.10 The percentage of *Aphis fabae* that flew for decreasing periods of time in mins (□) compared with the percentage that have 6, 8, 10 and 12 ovarioles (▧). (After Kennedy and Booth, 1963.)

investment in terms of ovariole number. The range and frequency distribution of reproductive investment is characteristic for a species, and even for a morph (Wellings *et al.*, 1980), and is another aspect of the polymorphism of aphids.

Thus aphids with small gonads have a greater urge to disperse, may fly for longer and more frequently, and are better able to survive starvation and to select host plants than aphids with large gonads. The result should be differerential dispersal. The long-distance dispersers should consist mainly of individuals with small gonads, and the flitters of individuals with large gonads. Plants differ in quality and distribution both within and between seasons. Therefore the degree of differential dispersal shown by an aphid clone will probably reflect both the within and between year heterogeneity of the environment. In favourable environments individuals with a high reproductive investment, the 'risk takers', will be fitter than those with a low reproductive investment, the 'risk averse', and vice versa (p. 84).

Timing

Daily patterns

By using suction traps, whose sampling efficiency is independent of wind speed and which segregate the aphids into hourly catches, Johnson and his colleagues (1957) revealed the factors that determine the number of *Aphis fabae* flying over a bean crop. It mainly depends on the rate of eclosion of the winged adults and the time it takes them to complete their teneral development. Flight is inhibited at night by low light intensity and sometimes during the day by low temperatures. The high rate of moulting in the early morning, followed by a lower rate during the rest of the day, sometimes rising again in the evening and falling almost to zero during the night, is possibly controlled by temperature. The teneral period between final ecdysis and first flight is temperature-dependent: aphids fly much sooner at higher temperatures.

Aphids become flight-mature all through the day and night, temperature permitting. When light intensity and temperature rise above the thresholds for take-off, the aphids that mature overnight, but are prevented from flying by cold and darkness, take off and give rise to the first flight peak of the day (Figure 7.11). The second flight peak later in the day represents the peak in number of newly moulted alatae that appear in the morning and complete their teneral development by the afternoon. The second flight peak declines in the evening because the temperature drops and light fades to intensities below the thresholds necessary for take-off.

The decline in aerial numbers at mid-day is due to a lack of flight-mature alatae and to the fact that most of the aphids in the air are newly flight-mature and fly at most for two hours. The model developed by Johnson *et al.* (1957)

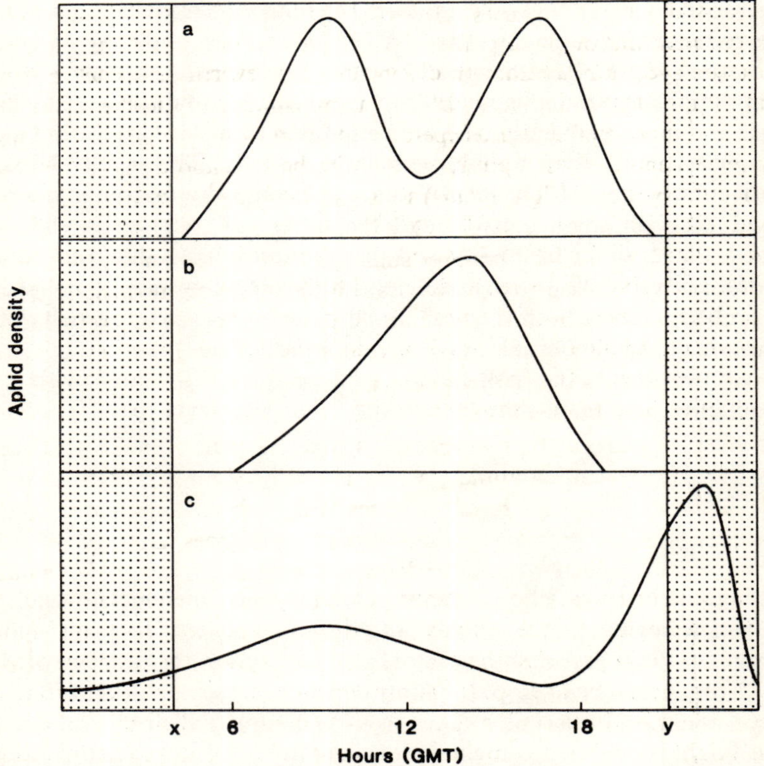

Figure 7.11 Daily flight activity of *Aphis fabae (a)*, *Eucallipterus tiliae (b)* and *Drepanosiphum platanoidis (c)* (X and Y time of sunrise and sunset, respectively).

accurately predicts the daily flight curves of *Aphis fabae*. Wind and humidity, which had previously been regarded as important factors in determining aphid flight, can be ignored when considering flight periodicity in the black bean aphid. In fact Johnson (1969) presents convincing evidence that most aphid migration occurs in windy weather. However, this is for aphids living on annual or short lived plants, which are often widely scattered and less apparent than trees. By riding the winds these species of aphids can scan large areas and so possibly improve their chances of finding other host plants.

Many species of aphids show the same bimodal flight periodicity as the black bean aphid, and the model developed by Johnson *et al.* (1957) also predicts the daily flight activity of the emigrant alatae of *Phorodon humuli* leaving hop (Taimr *et al.*, 1978). There would not be a bimodal flight periodicity if, after completing the teneral period, most alatae were able to fly for several days (Taylor, 1965). For example the lime aphid does not autolyse its wing muscles, is capable of flight throughout its adult life and has a single peak of flight activity just after midday. The timing of greatest activity in this case is possibly determined by the aphid's high light-intensity and temperature thresholds for flight (Figures 7.5, 7.11).

In contrast, although capable of flight throughout adult life, the second-generation sycamore aphids present in summer nevertheless show a bimodal flight periodicity with a peak early in the morning and another after sunset. The low light-intensity and temperature thresholds for flight in this species (Figure 7.5) enable it to fly early and late in the day, and the decline in aerial numbers at mid-day (Figure 7.11) is associated with high wind speeds, which inhibit take-off (Haine, 1955; Dixon and Mercer, 1983). The dramatic increase in flight activity around sunset, although dependent on the calm conditions that tend to prevail then, is heightened by an intrinsic urge to take off at this time of the day. However, the flight activity of first generation sycamore aphids in still conditions is strikingly different, as in this case most take off in the early afternoon when in the field wind speeds are likely to be high (Dixon and Mercer, 1983).

In still conditions second-generation sycamore aphids, possibly in common with other tree-dwelling aphids, can retain control of their flight paths and remain within sight and reach of trees. Flight in still conditions may be associated with the permanence and apparency of trees. Directed local flight may be the most effective means of dispersal from tree to tree within a habitat, referred to as trivial flight activity. The different, early afternoon, flight activity of first-generation sycamore aphids is associated with their greater tendency to disperse which possibly serves to transport the aphid further, to new habitats (woodlands).

Thus the form of the daily flight curves is dependent on the availability of flight-mature aphids, their urge to fly and their light intensity, temperature and wind-speed thresholds for take-off. However, only the study of other species will decide the extent to which one can generalize about the

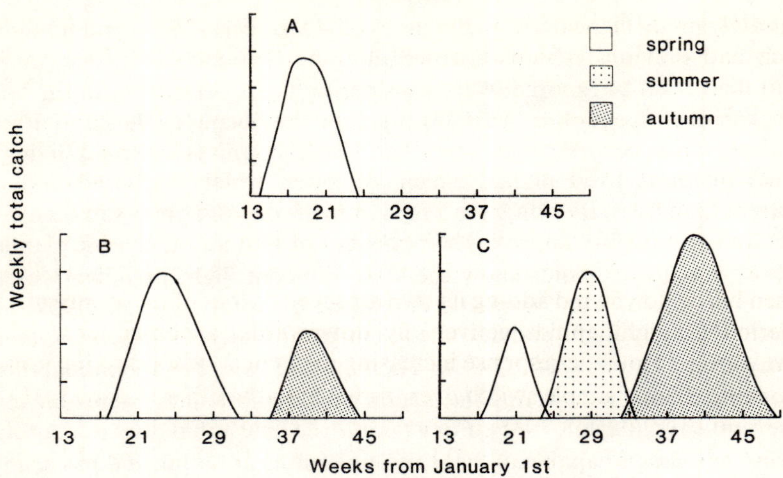

Figure 7.12 Seasonal flight activity of *Elatobium abietinum (A) Phorodon humuli (B)* and *Rhopalosiphum padi (C)*.

mechanisms underlying the different daily flight activity curves, their association with particular habitats and the evolutionary significance of the timing.

Seasonal patterns

Seasonal periodicity in the flight activity of aphids is also well marked. The green spruce aphid (*Elatobium abietinum*) has a single flight peak in May just as its host is breaking its buds; the host-alternating hop aphid (*Phorodon humuli*) has two peaks and the bird cherry-oat aphid (*Rhopalosiphum padi*) three (Figure 7.12). The spring and autumn peaks of flight activity in the two host-alternating aphids represent the populations leaving and returning to the primary hosts, respectively. In the bird cherry-oat aphid there is an additional period of flight activity during the summer when the aphid redistributes itself between its secondary host plants.

Thus for the host-alternating species the seasonal pattern of flight activity has a clear adaptive significance. In the autoecious green spruce aphid it serves to disperse the species. However, the significance of why dispersal in this and many other autoecious species of aphid should be confined to one period of the year, or of its timing, is unknown.

Distance travelled

Atmospheric circulation dominates the aerial movement of aphids because of their low flight speed, 1.6 to 3.2 km h^{-1}, relative to the speed of movement of air. In the very still conditions close to vegetation some aphids fly within sight of vegetation, alight at short intervals and do not fly very far. Taylor (1965, 1974) calls this 'boundary layer migration'.

Aphids that fly upwards and out of the boundary layer enter air masses that are likely to be moving. Up to $5\,\mathrm{km\,h^{-1}}$, air flow is laminar and under such conditions an aphid flies upwards, until there is a balance in the tendency to fly upwards in response to the light from the open sky and to fly downwards in response to the light reflected from vegetation (Kennedy, 1976). This height is a few metres, probably less than 8, from the ground (Taylor 1965). The aphid is then borne downwind adding its own air speed to that of the air mass. After a period of flight aphids actively fly downwards, a consequence of the downward phototactic response increasing in strength relative to the upward one, rather than a simple switch over from one to the other (Kennedy, 1976). This kind of migration is called 'stratiform' (Taylor, 1965).

On most days the average wind speed exceeds $5\,\mathrm{km\,h^{-1}}$, and flow is not laminar, because of either convection, or turbulence produced by surface roughness. The eddies produced vary in size from 15 m to 460–600 m in diameter, taking up to an hour to circulate. Aphids entering these air masses are borne along in a rolling motion until they actively fly down out of the air mass and into the boundary layer.

Support for the view that most aphid migration occurs by riding turbulent convection comes from their aerial vertical density profile. It is similar in form to that for dust and plant spores:

$$f(z) = c(z + ze)^{-\lambda}$$

where $f(z)$ is the number of aphids per unit volume (density) at height z; c, a scale factor related to the population size of the source; λ, gradient of density and ze, a measure of the departure from linearity on a double logarithmic scale. In the lower strata of the air at least, the gradient for dust (-0.28) is remarkably similar to that for aphids (-0.35). This led to the suggestion that the relationships for the deposition of dust and plant spores in relation to distance from source could be adapted to describe aphid dispersal.

This proved difficult to test because the aerial populations of most species of aphid are made up of individuals from many contiguous sources. However, in Britain the gynoparae and males of the host-alternating hop aphid come mainly from two discrete sources, the hop gardens in Hereford and in Kent, which are 180 km apart (Figure 7.13). Using seven years of census data collected by the Rothamsted Insect Survey, Taylor *et al.* (1979) produced density profiles radiating out from each of the sources, showing that the migration is not directional. As the sources reached similar densities the two relationships between density and distance were combined (Figure 7.14). This elegantly revealed that the aphid aerial distribution curve is similar to that expected if the aphids were simply transported on the wind with the gradient dependent on the degree of turbulence rather than by random diffusion.

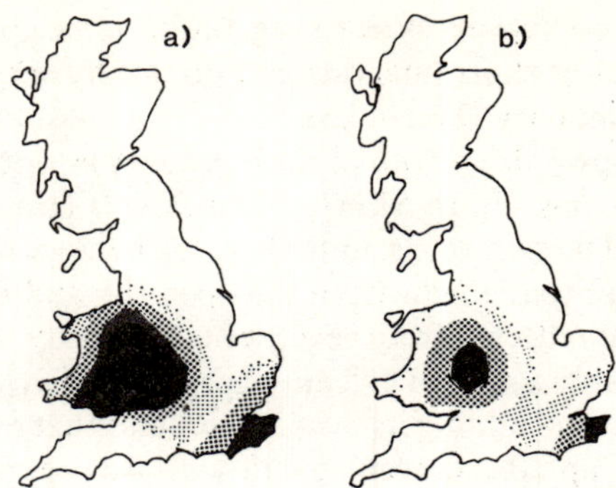

Figure 7.13 Mean aerial distribution of sexuals of the hop aphid, *Phorodon humuli*, for 1971 to 1976. *a*, gynoparae, *b*, males. The concentrations are over the Hereford and Kent hop gardens. (After Taylor *et al.*, 1979.)

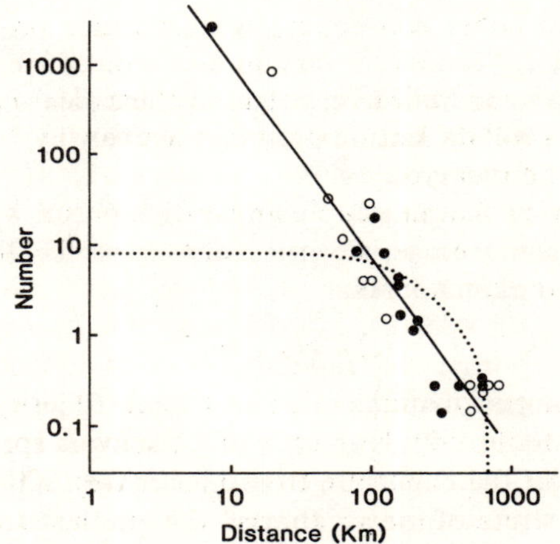

Figure 7.14 The average number of gynoparae and males of the hop aphid caught from 1971 to 1977 in suction traps at varying distances from the Hereford (●) and Kent (○) hop gardens (cf. Figure 7.13).—— deposition of wind-borne fine soil particles, ... expected from random diffusion. (After Taylor *et al.*, 1979.)

An aphid's migratory tendency, defined as its tendency to fly upwards, is also likely to determine the form of the density distance function. However, it is unknown what selection pressures have shaped the frequency distribution of the distances travelled by individuals, in terms of short and long flights, that is to say, the migratory ambit of a species, although the actual distances travelled are determined by wind speed (Taylor *et al.*, 1979). It is likely that the migratory ambit is linked with reproductive investment so that those individuals with the fewest ovarioles and greatest migratory urge travel furthest. That the hop aphid, and many others, show differential reproductive investment lends support to this idea (Figure 7.15).

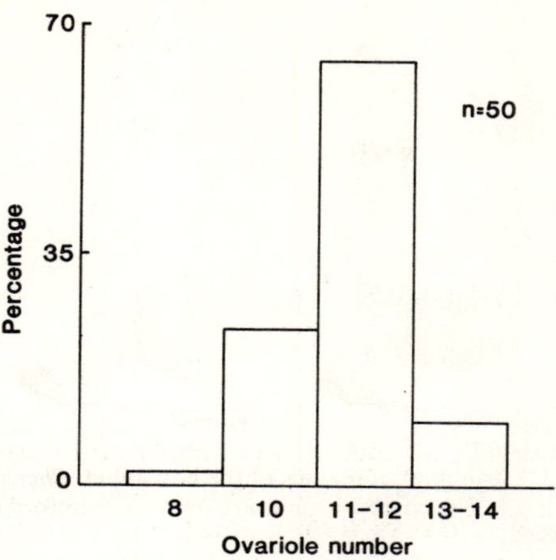

Figure 7.15 Percentage of emigrant hop aphids that have 8, 10, 11–12 and 13–14 ovarioles.

The median distance travelled by hop aphid sexuales is about 15–20 km, with a 95% limit of 100–150 km. The direct interchange of sexuales between the two hop centres is therefore minimal in a season, but inevitable over a series of years, as the aphid can survive on the widespread winter host and on wild hops in surrounding areas, albeit at a low density (Taylor *et al.*, 1979).

Aphids lifted by convection high enough to reach the persistent high-altitude (300–900 m) air streams, called low-level jet streams by meteorologists, could be transported very rapidly over great distances. However, an aphid would have to keep flying in order to stay in the jet stream as it would experience little convective lift. It is likely that the living spruce aphids Elton (1925) found on a glacier in Spitsbergen had made the 1300 km journey from the Kola peninsula in this way, flying for up to 24 h in the long summer days of the northern latitudes. As Wellington (1983) stressed, although they are weak fliers, aphids and other insects can disperse over a wide area by exploiting the ephemeral but very structured nature of air movement.

Summarizing, the tendency to disperse is an evolved adaptation (Johnson, 1969) that has enabled aphids to spread the chance of survival in space, and to seek out and colonize plants that are above average quality. Although the speed and direction of flight is usually governed by the wind, aphids nevertheless terminate a flight by actively flying downwards and settling, and thus they have some control over the distance that they travel. The exceptional very long-distance migrations are achieved by riding the high-altitude jet streams. The patterns of daily and seasonal flight activity are shaped by the availability of flight-mature aphids and their light intensity, temperature and wind-speed thresholds for flight. However, except for the spring and autumn flights of host-alternating aphids, the significance of the timing of the flights is poorly understood.

8 Population dynamics

Aphids can become very abundant, noticeably so on agricultural crops. A hectare of field beans can produce 4000 million alatae of the black bean aphid (*Aphis fabae*) (Way and Banks, 1967), and the cereal aphid, *Metopolophium dirhodum*, can achieve population densities of 220 per tiller or 1000 million per hectare of wheat. Although aphids are small these numbers are equivalent in mass to an elephant and a cow respectively. Insect predators feeding on these aphids can also reach impressive numbers. In England in 1976 280 000 seven-spot coccinellid adults, *Coccinella septempunctata*, matured per hectare of wheat after feeding on the cereal aphid *Sitobion avenae*, which nevertheless achieved a peak population of 200 million per hectare. The estimated 24 000 million seven-spot ladybird beetles produced in the county of Norfolk combined with some produced in other counties in 1976 gave rise to a ladybird 'plague' that drove people off the beaches in parts of East Anglia.

Populations have either been studied on a few plants in a relatively small area or, uniquely, in the air over the British Isles. The often detailed studies of aphids on plants only cover a minute fraction of the population and range of these highly mobile insects. However, by combining the results of the two approaches it is possible to extrapolate from the understanding of what determines the numbers of aphids on a few plants to understanding the population dynamics of the species over a wide area, and so consider their spatial dynamics (Dixon, 1979).

Intensive studies

The dynamics of aphid populations, with their overlapping generations and unstable age structures, are not easily studied using explicit formulae as in analytical models. Most of our ideas about aphid population dynamics have been derived from detailed, often lengthy simulation models consisting of a specified routine of arithmetic operations. Although few aphid populations have been studied for any length of time, nevertheless population studies of aphids on both herbaceous and woody hosts are beginning to reveal patterns in the changes in abundance from year to year and we have some understanding of the underlying processes.

Patterns

Aphids that do not go through a sexual cycle, but overwinter viviparously, are anholocyclic (p. 61). Although they are often well protected against cold conditions, they are nevertheless likely to suffer a much higher mortality in severe winters than are holocyclic species overwintering as eggs, and this possibly accounts for the difference in the patterns of changes in aphid numbers from year to year between anholocyclic and holocyclic species, at least in temperate regions.

Anholocyclic species. Outbreaks of the spruce aphid (*Elatobium abietinum*) are associated with mild winters (Bejer-Petersen, 1962; Carter, 1972; Powell and Parry, 1976). When temperatures fall to – 7° C or below, ice forming inside the needles of the host seeds ice crystals in the attached aphid and kills it (Powell, 1974). However, the waxy bloom of this aphid that serves as an anti-wetting layer may hold off ice crystals and, in preventing them from contacting the aphid's body, delay fatal inoculative freezing (Bevan and Carter, 1980). Although mortality is catastrophic as a result of freezing, the aphid also steadily decreases in numbers once the average weekly temperature falls below 4° C (Figure 8.1*A*). However, in springs following mild winters there are often large numbers of overwintered aphids that have survived on spruce. Just prior to bud burst the aphids multiply rapidly and, if numerous to begin with, will achieve outbreak levels (Figure 8.1*B*) and cause defoliation.

The number of cereal aphids (*Sitobion avenae*) flying in parts of England in early summer is also correlated with the severity of the previous winter (Figure 8.2). It is likely that the aphid survives well and possibly even increases

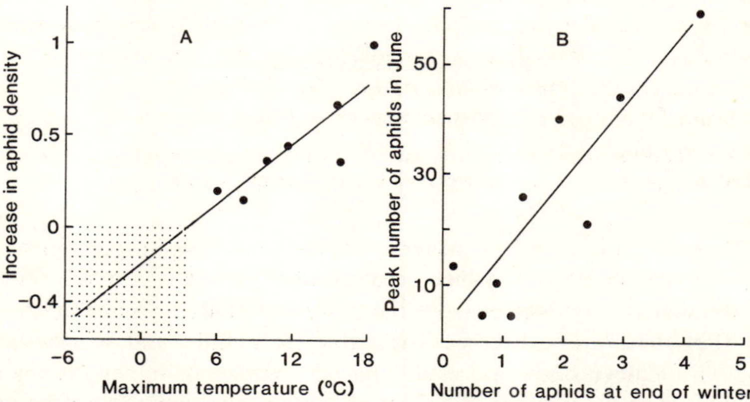

Figure 8.1 Factors determining the abundance of the green spruce aphid: *A*—the increase in density per week, measured as the logarithm of the number of aphids per gram dry weight of needles, in relation to the maximum temperature experienced each week; *B*—peak number of aphids per 100 needles in June in relation to the number of aphids per 100 needles present at the end of winter. Stippling, negative rates of increase. (After Powell and Parry, 1976)

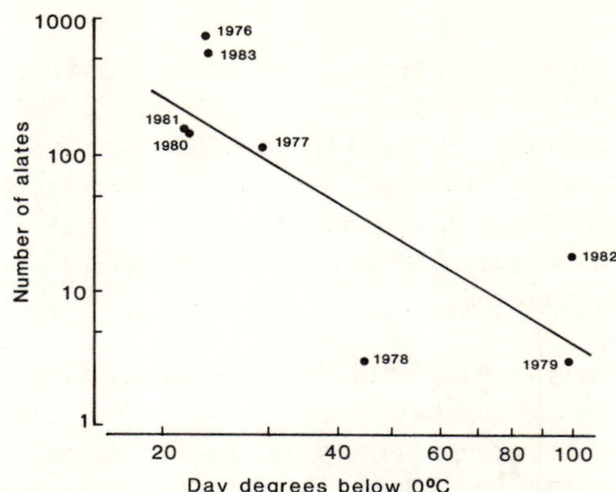

Figure 8.2 The number of alates of *Sitobion avenae* caught in the Broom's Barn suction trap before the end of flowering of wheat in the field, from 1976 to 1983 in relation to the number of day degrees below 0° C in the period October to April of the previous winter.

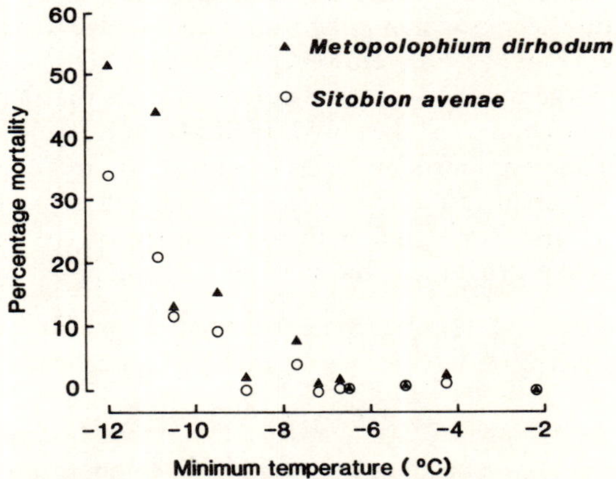

Figure 8.3 The percentage mortality over 4 days of two cereal aphids in relation to the minimum temperature experienced during each period. (After Williams, 1980.)

in mild winters (Figure 8.3). Further increases in numbers in spring give rise to numerous alatae, many of which colonize cereal crops. There is a relatively fixed number of day degrees from the time the aphids colonize the crops to when they mature and become unsuitable for aphids, and so whether the aphid will achieve outbreak levels and do significant damage is mainly a consequence of the number of aphids arriving in the crop. Therefore, where the aphid mainly overwinters viviparously the severity of winter is important in determining fluctuations in abundance from year to year.

The green peach aphid (*Myzus persicae*) is an important vector of beet yellows virus, and is mainly anholocyclic in England. The incidence of beet

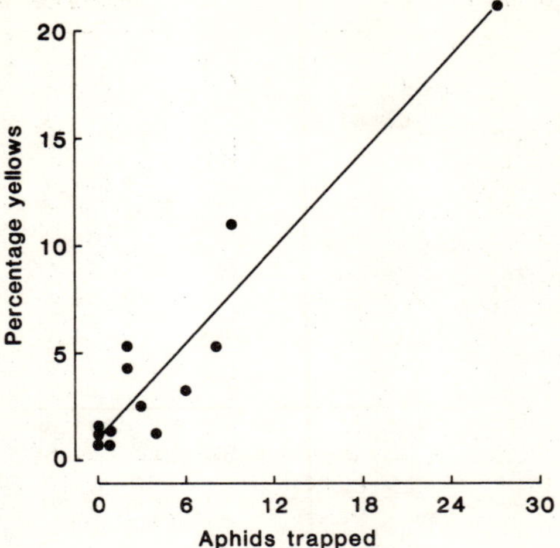

Figure 8.4 Percentage of sugar-beet plants with yellows in relation to the number of *Myzus persicae* caught on sticky traps in May and June. (After Heathcote, 1974.)

yellows virus in sugar beet is closely associated with, first, the number of days in the first three months of the year when the temperature falls below 0° C, and the mean weekly temperature in April. Second, the incidence appears to be determined by the numbers of *M. persicae* flying in May and June (Figure 8.4). The temperature in late winter determines how many aphids survive, and the rate of increase of the survivors is determined by the temperature in April. When the aphid is abundant in spring sugar beet crops are at risk (Figure 8.4). Thus using winter and spring temperatures it is possible to predict the incidence of beet yellows virus very accurately in a particular year (Watson *et al.*, 1975).

To overwinter viviparously is hazardous for aphids because their rate of increase and their survival are very dependent on temperature. Cold winters result in a rapid decline in numbers even to the point of extinction, whereas in mild winters numbers increase. Thus winter conditions are very important in determining the year to year fluctuations in the abundance of anholocyclic species.

Holocyclic species. Although egg mortality can be very high (reviewed by Leather, 1981) there is no evidence that it is highest in very severe winters. Mortality occurs at a fairly constant rate throughout winter (Figure 8.5). (Leather, 1980*b*; Thornback, 1983). The more eggs that are laid, the more survive to hatch, but it is not known whether the proportion surviving is independent of density. Thus, winter conditions are not as important, in determining the year to year fluctuations in abundance, in holocyclic species as they are in anholocyclic species.

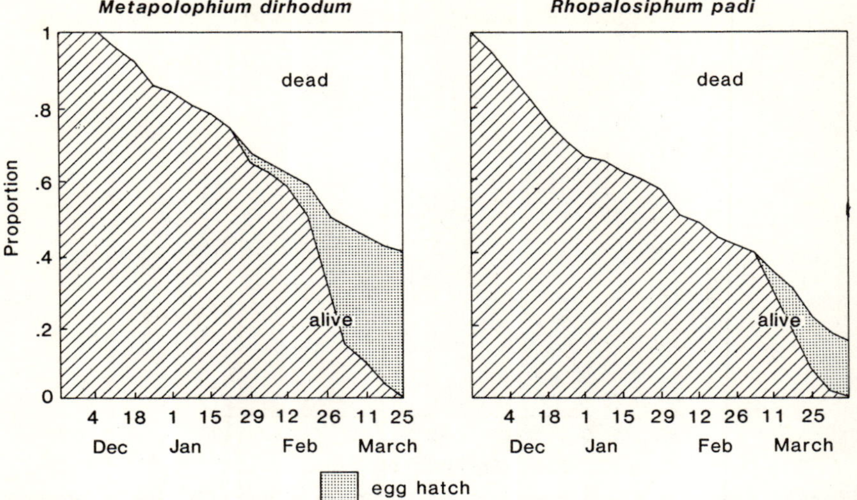

Figure 8.5 Proportion of the eggs of *Metopolophium dirhodum* and *Rhopalosiphum padi* surviving over winter. (After Leather, 1980*a*, and Thornback, 1983.)

For the few species of holocyclic aphids that have been studied in detail a high level of abundance in spring is followed by low numbers in autumn and vice versa. This seesaw pattern has been observed in one species of host alternating aphid and three species of autoecious tree-dwelling aphids.

The black bean aphid (*Aphis fabae*) overwinters on spindle (*Euonymus europaeus*) (Figure 6.2); leaves spindle in June to colonize broad beans (*Vicia faba*), sugar beet (*Beta vulgaris*), and various other herbaceous plants (Jones, 1942*b*; Way, 1971); and returns to spindle in autumn. The more aphids on spindle in autumn, the more eggs are laid. More than 60% of these eggs hatch the following spring (Way and Banks, 1964). The greater the number of aphids hatching in spring the greater the number of summer migrants (Behrendt, 1966, 1969; Way and Cammell in Crawley, 1983). Therefore, although predators and parasites can inflict heavy mortality on the aphid, both on spindle and broad bean (Banks, 1955; Way and Banks, 1968), the numbers of the aphid are not regulated in this way. Up to this point in the life cycle of *A. fabae* more aphids leave each stage than enter it. However, there is an inverse relationship between the abundance of summer migrants that leave beans and autumn migrants that return to spindle (Figure 8.6) (Way, 1967). This results in a tendency for aphid outbreaks to occur in alternate years (Behrendt, 1966; 1969; 1971; Muller 1966*b*; Way, 1967; Way and Cammell, 1973). Weather, through its effect on the aphid's rate of increase and colonization of plants, can modify this pattern as can the activity of natural enemies (Way, 1967).

Similarly in the lime aphid (*Eucallipterus tiliae*) and sycamore aphid (*Drepanosiphum platanoidis*) there is an inverse relationship between spring

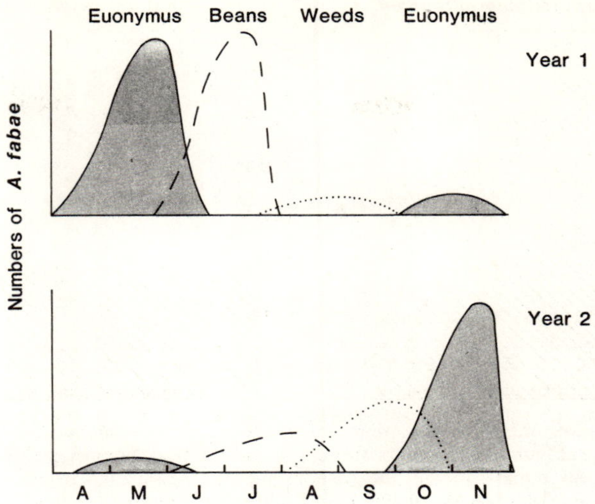

Figure 8.6 Diagrams of seasonal and annual changes in *Aphis fabae* populations on spindle (*Euonymus*) (stippled), field beans (– – –) and weeds (. . . .) showing a typical 2-year cycle with alternating large and small May-July populations. (After Way, 1967.)

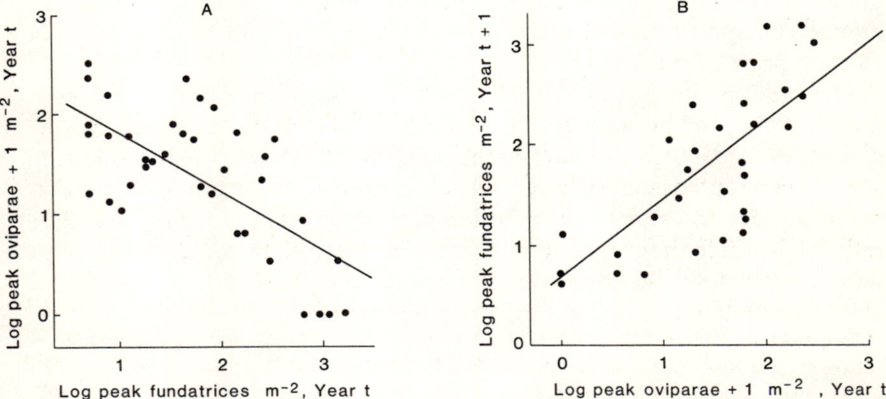

Figure 8.7 The peak number of oviparae of the lime aphid (*Eucallipterus tiliae*) in relation to the peak number of fundatrices each year (*A*) and the peak number of fundatrices in relation to the peak number of oviparae the previous year (*B*). Each point represents 1 tree in 1 year.

and autumn numbers, and a positive relationship between autumn and spring numbers (Figures 8.7, 8.8).

Although there are several parthenogenetic generations between egg-hatch in spring and egg-laying in autumn, the ratio of the densities of the aphids at these two stages gives a measure of the overall rate of increase of the population. Thus, in the analysis of changes in lime and sycamore aphid

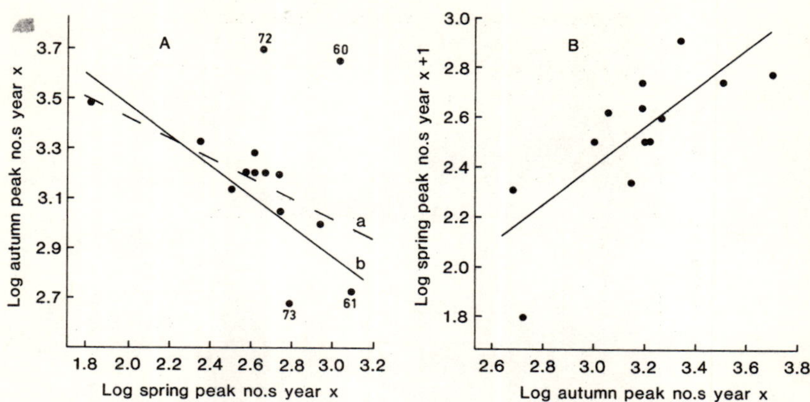

Figure 8.8 The peak number of sycamore aphids (*Drepanosiphum platanoidis*) in autumn in relation to the peak number present in spring (*A*) and the peak number present in spring in relation to the peak number present the previous autumn (*B*) for each year from 1960–1973. (Regression line *a* in *A*, $y = 4.65 - 0.59x$, $r = 0.81$, $P < 0.01$, does not include 1960 and 1972; *b*, $y = 4.25 - 0.41x$, $r = 0.91$, $P < 0.01$, does not include 1960, 1961, 1972 and 1973. (After Dixon, 1979.)

numbers a meaningful first step is to determine the relationship between the number of aphids present after egg-hatch in spring and those present in autumn prior to egg-laying.

The equation expressing the relationships between the number of fundatrices in spring and the number of oviparae in autumn of the same year, and the number of oviparae in autumn and the number of fundatrices in the following year, refer to aphids that survived. On a logarithmic scale a population grows from its previous value by the addition of the logarithm of the reproductive rate (R), which in the case of aphids is the rate achieved by the clones during a season. Therefore, the expected relationship, e.g., between the number of oviparae (O) in autumn and fundatrices (F) in spring is:

$$\log O = \log F + \log R$$

Mortality related to the number of fundatrices is represented by a negative power of the number of fundatrices F^{-x}, and the relationship becomes:

$$\log O = \log F + \log R - x \log F$$

The value of x represents the degree of density dependence.

In both the lime and sycamore aphids the value of x is approximately 1.5, which indicates that there is an overcompensated density-dependent factor acting within years. Between year mortality is inversely density-dependent (Dixon, 1970*b*, 1971*d*, 1979).

Wind is an important disturbing factor through its effect on mortality, both in the lime and sycamore aphids, and temperature causes fluctuations through

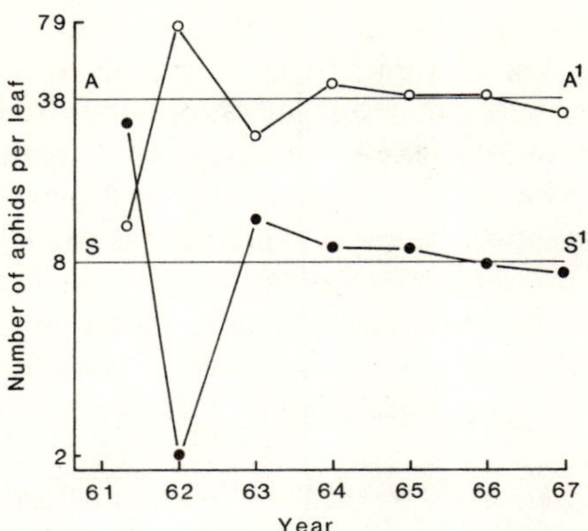

Figure 8.9 Peak number of sycamore aphids per leaf in the spring (●) and autumn (○) from 1961 to 1967. A–A[1] and S–S[1] predicted equilibrium densities for autumn and spring, respectively. (After, Dixon, 1970*b*.)

its effects on flight in the lime aphid (Dixon, 1979; Barlow and Dixon, 1980). The particularly still autumns of 1960 and 1972 resulted in the highest autumnal populations observed for the sycamore aphid (Figure 8.8). Following these autumns the action of the overcompensated density-dependent factor led to fluctuations in numbers, but the amplitude of the fluctuations decreases with time (Figure 8.9).

Processes

The black bean aphid (Aphis fabae). Understanding the processes that give rise to the inverse relationship between the abundance of summer migrants and autumn migrants (Way, 1967) will clarify what regulates the numbers of this aphid. There is no detailed information on the processes, but it is suggested that regulation is due primarily to insect predation, in particular that inflicted by coccinellid beetles. The large number of coccinellids nourished by the abundant aphid population on secondary hosts in early summer is thought to be able to reduce the numbers of aphids that colonize other plants (e.g., *Chenopodium album*) in late summer. These same predators may overwinter and also prevent the production of large numbers of migrants from spindle the following spring. This probably accounts for aphid outbreaks in alternate years as described by Behrendt (1966, 1969, 1971), Müller (1966*b*), Way (1967) and Way and Cammell (1973). Furthermore, weather affects the system through its effect on the aphid's rate of increase, successful colonization of plants and the activity of its predators (Way, 1967).

However, the fate of the offspring of the summer migrants has not been investigated. In years when they are produced in large numbers, summer migrants must face a shortage of suitable host plants. This could lead to intense intraspecific competition and so to the production of few autumn migrants. Warm dry years that further favour the production of summer migrants are unfavourable for secondary host plants like *C. album* (Weismann, 1967). The scarcity of secondary host plants late in the season must further intensify intraspecific competition. Because of their low fecundity (Dixon and Wratten, 1971) many of the small summer migrants, produced when the aphid is abundant on beans (Way and Banks, 1967; Way, 1968), may have difficulty in establishing colonies on other host plants in later summer. The relative importance of these factors is unknown.

The lime and sycamore aphids (Eucallipterus tiliae *and* Drepanosiphum platanoidis). The regulatory mechanisms operating in these two species have been determined by studying laboratory populations and by field experiments. Seasonal changes in the abundance of these two species have been followed for 9 and 15 years, respectively.

Even in the absence of natural enemies laboratory populations of the *lime aphid* suffer a sudden decline in numbers in June or July if the numbers are high at the beginning of the season (Dixon, 1971*d*). This gives an inverse relationship between initial and final numbers within a year, similar to that seen in the field (Figure 8.12*A*). Although overcompensation occurs in cage populations it is less marked than in the field. This reduced overcompensation, together with generally greater numbers of oviparae in cage populations, may be related to the absence of factors such as wind and predation.

Given a knowledge of the separate component processes, a simulation model of their combined action yields the following hypothesis that accounts for behaviour of the lime aphid populations. Weather is important as a disturbing factor and as a major determinant of peak numbers, and numbers of fundatrices in any one year. The population is regulated by a hierarchy of processes controlling population growth at different densities and capable of being substituted one for another. The ultimate limit to growth in any one season does not appear to be available space, or death of the host through over-exploitation, but rather a decline in its quality to a level at which aphid survival is greatly reduced. Below this level aphid numbers are regulated by a combination of an increase in the numbers of the two-spot coccinellid in response to aphid densities early in the year and, at higher densities, proportionately more adult aphids dispersing by flight, which depends on densities experienced during nymphal development. The role of these various processes relative to the within year dynamics of the aphid are summarized in Figure 8.10 (Barlow and Dixon, 1980). Thus as in the thimbleberry aphid (*Masonaphis maxima* (Mason)) there is no single 'key factor' (Gilbert, 1980) regulating lime aphid abundance.

This interpretation rests on the existence of a cumulative or integrated density-dependent effect. The weights of individual aphids developing either

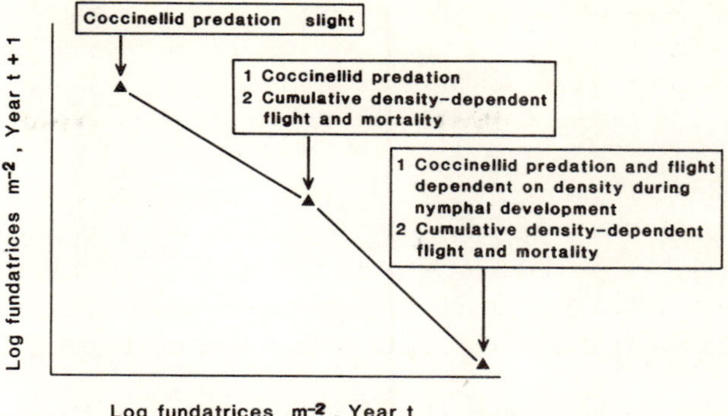

Figure 8.10 A summary of the main factors that regulate lime aphid numbers in years of low, medium and high numbers of fundatrices and the resultant number of fundatrices the following year. (After Barlow and Dixon, 1980.)

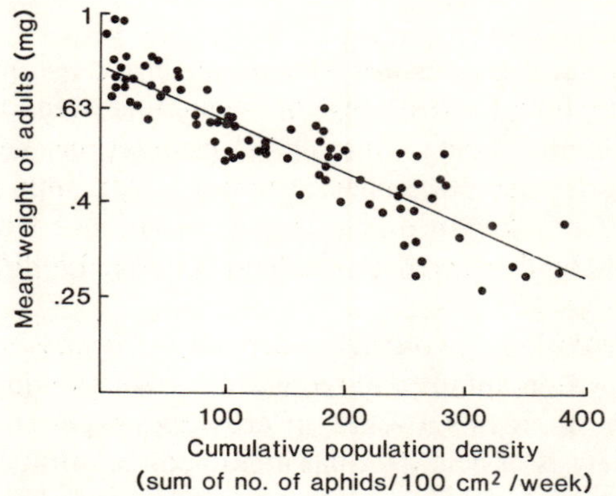

Figure 8.11 Mean weight of adult lime aphids from the field in relation to the average cumulative population density experienced by the leaves on which the aphids developed.

in the field or laboratory are more closely associated with the cumulative density of aphids that has fed on the leaves up to that time (Figure 8.11), than the density experienced by an aphid during its development. In addition, in the field the mortality of adults and incidence of flight increases strikingly after prolonged heavy infestation of the trees, corresponding to a cumulative density of about 250 aphid-weeks/100 cm^2 of leaf surface. This could be due to the transmission of crowding effects from generation to generation, possibly through birth weight, or from changes in the quality of the host plant in response to aphid attack, or both. Aphid induced changes in the host plant have been shown experimentally to affect the time aphids take to reach maturity and adult weight, but do not affect aphid mortality or flight activity.

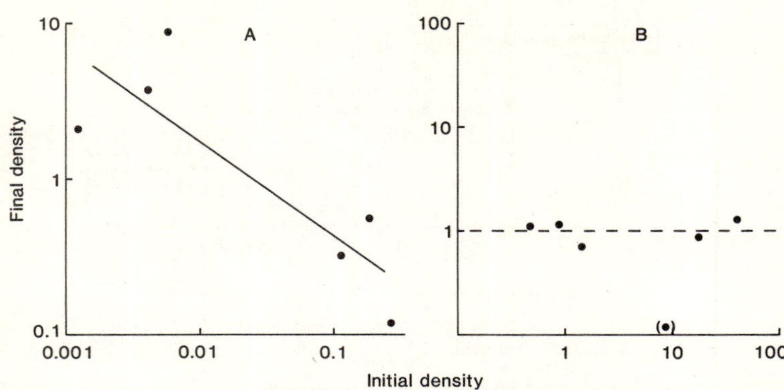

Figure 8.12 The number of lime aphid eggs laid on the stems of lime saplings in relation to the number of aphids initiating the population at the beginning of the year (*A*) and peak number of sycamore aphids on sycamore saplings in relation to the number of aphids per unit area on the buds at the beginning of the year (*B*) (Lime aphid numbers are per square centimetre of leaf, sycamore aphid numbers per square centimetre of bud surface and leaf, respectively, (●)sapling died at end of year). *A* after Brown, 1975; *B* after Chambers, 1979.

The maximum numbers reached in autumn in a cage population of the *sycamore aphid* are similar, irrespective of the number of fundatrices initiating the populations in the spring. Overall the populations behave in a compensatory density-dependent manner, attaining a maximum autumnal population of 1 aphid/cm^2 (Figure 8.12). This is at variance with the overcompensatory relationship observed in the field (Dixon, 1970*b*; Wellings *et al.*, 1985).

In the cage populations, two density-dependent processes operate: firstly, flight of first generation aphids and secondly, a reduction in the fecundity of second generation aphids as a result of crowding experienced during their development. Analysis of field data has identified the duration of aestivation of the second generation as being density-dependent, and has also shown that reproduction in autumn depends on the cumulative effect of densities in spring and summer. All these processes are density-dependent in an undercompensatory fashion, and, as a result, cannot individually account for the population behaviour observed in the field.

However, weather, in particular wind, can have a dramatic effect on sycamore aphid population dynamics (Dixon, 1979). Wind causes leaves to brush together, which dislodges the aphids, most of which are then lost to the population (Dixon and McKay, 1970). Thus the amount of space available for sycamore aphids may be limited to a small fraction of the total leaf area, especially during periods of high winds. The leaves in fact can act as devices converting weather into aphid mortality. In the exceptional still autumns aphids are able to achieve most of their potential rate of increase, and however few there are at the beginning of autumn, there is sufficient time to reach saturation levels of 1 aphid/cm^2 before the end of the season. In the more

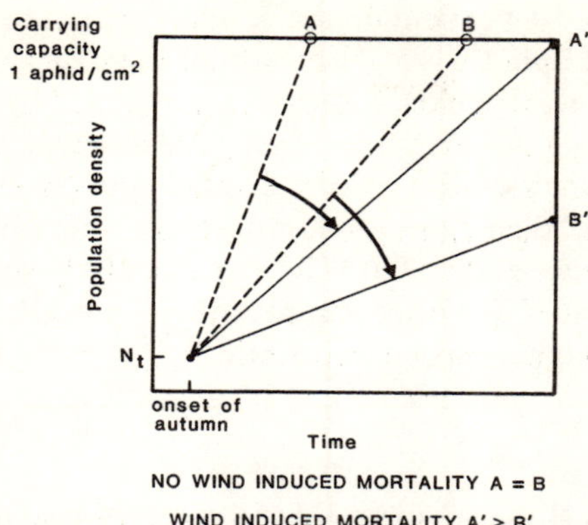

Figure 8.13 The effect of wind-induced mortality (→) on the peak autumnal numbers achieved by sycamore aphid populations with a high (*A*) or low (*B*) potential rate of increase. Slopes of the lines represent the realized rates of increase.

usual windy autumns the aphid only achieves a fraction of its potential rate of increase, and the abundance achieved will reflect the potential rate of increase: low in years when the aphid is abundant in spring and summer and vice versa (Figure 8.13) (Chambers *et al.*, 1985).

Natural enemies

There is no doubt that the natural enemies of aphids can reduce their rate of increase, occasionally dramatically, and the use of hymenopterous parasites in biological control of a few aphid pests would appear to have been successful. However, the precise role of natural enemies will remain uncertain and subject to speculation until their effect is quantified. For example, the critical period in the regulation of the numbers of the black bean aphid is between leaving beans in summer and returning to spindle in autumn. Although large numbers of ladybird beetles are often present in years when the aphid is abundant on beans the role of these predators in suppressing the aphid's numbers on other secondary host plants in late summer has not been studied.

Because the abundance of a particular species of aphid can change dramatically from year to year aphid populations need to be followed in detail for several years before and after the introduction of a biological control agent to be sure that the introduction results in a real change in the average level of abundance. This is rarely possible. Even if it were, the role of a hymenopterous parasite, like *Trioxys pallidus*, in reducing the abundance of the walnut aphid in California (van den Bosch *et al.*, 1979) can only be determined if the parasite is seen as part of a system, which includes the interaction between the host plant and the aphid as well as other natural enemies and hyperparasites, rather than a simple aphid/parasite interaction (Dixon, 1977). The successful use of hymenopterous parasites to reduce the abundance of the alfalfa aphid may have also been largely dependent on the development of resistant cultivars of alfalfa on which the aphid has a lower rate of increase (van den Bosch and Messenger, 1973). Cereal aphids, similarly, are attacked by an array of aphid-specific and polyphagous predators and parasites which in certain years appear to have a dramatic effect on cereal aphid abundance (Chambers *et al.*, 1982). However, until their effect is quantified their role will remain uncertain (Chambers and Sunderland, 1983). This is particularly so because in killing aphids, natural enemies can often improve the conditions for the survivors, which then have a higher reproductive rate.

Spatial dynamics

Species like the elderberry aphid (*Aphis sambuci*) have an unstable spatial pattern of distribution, apparently occupying a dramatically varying fraction of the potential distribution in any one year (Figure 8.14). Elderberry, the winter host of this holocyclic species, is ubiquitous throughout Britain and so too are the summer hosts, various species of dock (*Rumex* spp). In contrast the whole of Britain is almost uniformly occupied by species such as *Rhopalosiphum insertum* and *Cryptomyzus galeopsidis*, but *Aphis corniella*, not surprisingly, is restricted to parts of England and Wales where its host, dogwood (*Cornus sanguinea*) occurs, and to SW Scotland where dogwood is grown as a garden shrub (Figure 8.15). Although the maps shown in Figure 8.14 do not provide direct evidence of movement between areas, they have been offered, along with others, in support of the contention that these changing distributions involve movement as well as mortality (Taylor and Taylor, 1979; Taylor, 1979).

It is now well established that the spatial variability in population abundance and average population density of a species are related in both

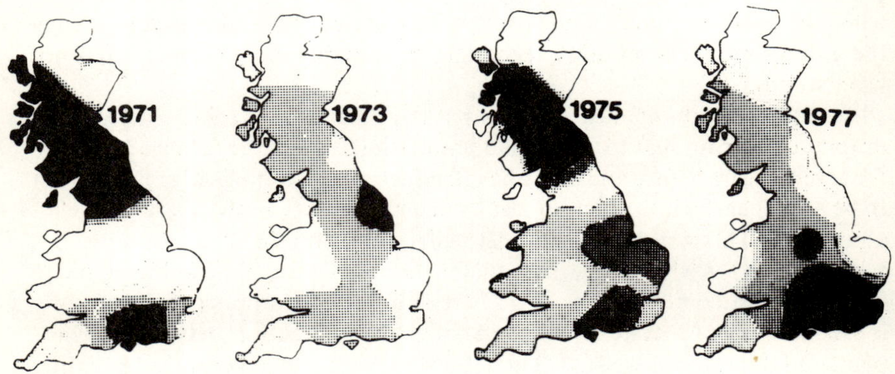

Figure 8.14 Maps showing changes in the aerial distribution of the elder aphid, *Aphis sambuci*. (After Taylor and Taylor, 1979.)

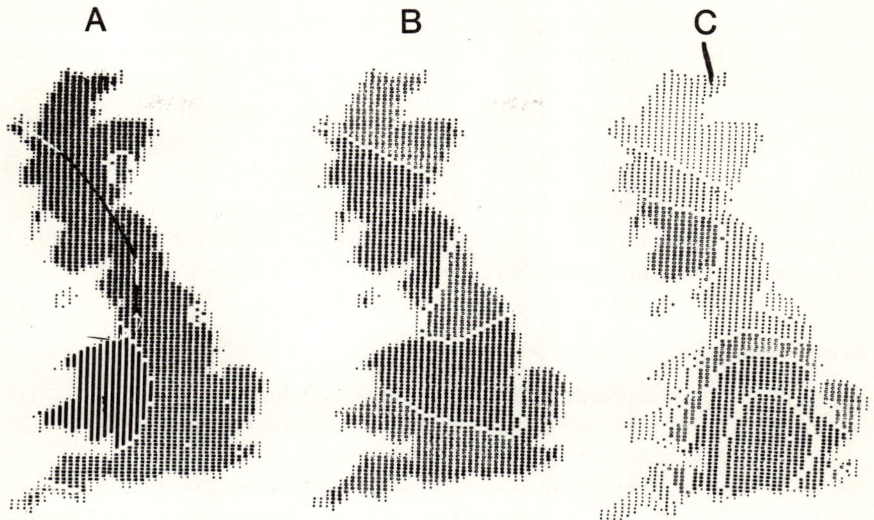

Figure 8.15 Maps showing aerial distribution of *Rhopalosiphum insertum (A), Cryptomyzus galeopsidis (B)* and *Aphis corniella (C)*. (After Taylor, 1973.)

space and time. For a wide range of plants and animals the relationship between variance (V) and mean (M) conforms to a simple power law:

$$V = aM^b$$

where a and b are constants (Figure 8.16) (Taylor, 1961). The slope of the relationship (b) generally falls within the range 1 to 2, is a species-specific characteristic and is thought to reflect the balance (Δ) between the opposing behavioural tendencies to aggregate within ($H\rho^q$) and migrate from ($G\rho^p$), centres of population density:

$$\Delta = G\rho^p - H\rho^q$$

where the exponents p and q are rate constants for density-dependent migration and congregation, respectively, G and H are constants of proportion and ρ is population density (Taylor and Taylor, 1977, 1978). Simulation studies based on this model yield censuses that satisfy $V = aM^b$. That migration between patches is rarely random is deduced from the slope (b) of the relationship between the variance and the sample mean, which is greater than 1, and non-random dispersal is also proposed on evolutionary grounds (Taylor *et al.*, 1978; Taylor *et al.*, 1983).

This interpretation has been challenged by Hanski (1980, 1982) on the grounds that it is unreasonable to expect insects, especially aphids, to move between population sites in order to maximize individual fitness, particularly in an area as large as Britain. Hanski also re-analysed the results of Taylor

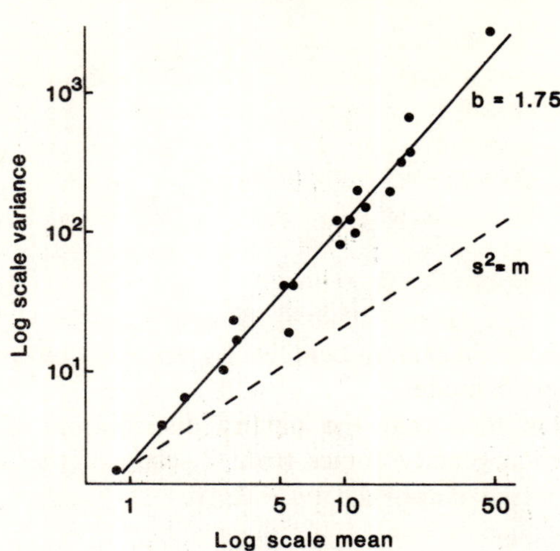

Figure 8.16 Logarithm of the variance of the distribution of sycamore aphids between leaves in relation to the logarithm of the mean number per leaf for each sample taken at weekly intervals throughout a year.

et al. (1980) to indicate that a value of *b* greater than unity does not necessarily imply density-dependent movement. Like Taylor and Taylor (1977), Hanski also argues that the value of *b* may reflect the ecology of species, but in a different sense. A specialist is only present in some habitats, even when very abundant, whereas a generalist possibly increases faster in the poorer habitats because the preferred habitats are fully occupied. Thus the slopes of the relationships between variance and sample mean for different habitats for specialists should have values of *b* greater than 2, and generalists, values less than 2. This is not supported by the values of *b* for 97 species of aphids cited by Taylor *et al.* (1980). Specialist species of aphids, that live mainly on only one species of plant, and generalist species are very similar in both their range in values of *b* and the proportion of species in the two categories that have values greater than 2. The two most polyphagous (generalist) species cited, *Myzus persicae* and *Macrosiphum euphorbiae* have values of 2.63 and 2.25, respectively. However, as suction-trap catches contain aphids from all the habitats in the vicinity of the trap, they provide too coarse a method of sampling to test Hanski's idea.

A further theoretical consideration of this problem has shown that the relationship between variability and average abundance could be a consequence of chance demographic events in the dynamics of population growth and decline, in which migration is a random process. The form of these relationships is determined by the relative magnitude of birth, death, immigration and emigration rates that govern the dynamics of population change, and by the degree of spatial and temporal heterogeneity. When density-dependent factors are of limited significance, as in species that are *r*-strategists, chance variations in birth and death rates alone are sufficient to account for the relationship between variance and mean abundance and for the slope lying on average between 1 and 2. For *K*-selected species, in which strong density-dependent factors are assumed to be acting, a high degree of environmental heterogeneity additionally ensure that such relationships remain approximately linear. Therefore there would appear to be no need to invoke any complex behavioural mechanisms to explain the observed patterns (Anderson *et al.*, 1982). However, Taylor *et al.* (1983) justifiably argue that the simulations of Anderson *et al.* fit field results less well than simulations based on the behavioural Δ model.

Support for the idea that the patchy distribution of aphids reflects environmental heterogeneity comes from a study of the mechanisms that govern the distribution of aphids between leaves (cf. Figure 8.16). Periodically aphids cease feeding, move off and colonize another leaf at random. The length of time for which an aphid stays on a leaf depends on its temperature, exposure to the sun, frequency with which its underside is brushed by other leaves (Figure 8.17), its nutritive status and the presence of other aphids (Dixon, 1970*a*; Dixon and McKay, 1970; Dixon and Mercer, 1983). Thus kinetic movements in response to the variable quality of the environment result in this aphid aggregating and reproducing mainly on the relatively few leaves that have a favourable combination of microclimate, degree of infestation and nutritive status.

A mild winter or calm autumn results in dramatic increases in the

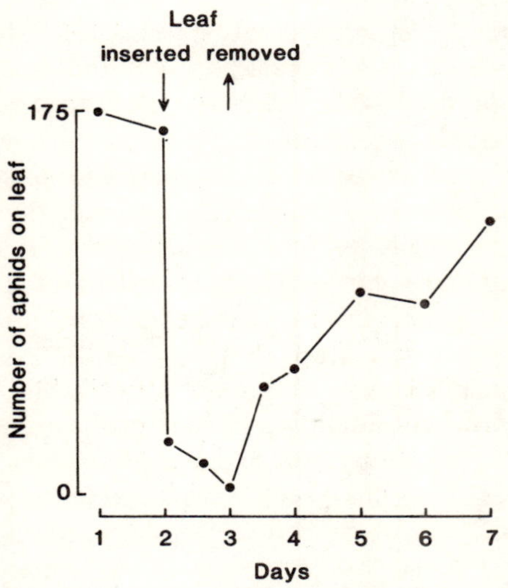

Figure 8.17 The effect on the number of aphids on a leaf of inserting another leaf immediately below it.

abundance of cereal and sycamore aphids, respectively, and fluctuations in their abundance are reflected in suction-trap catches (Dixon, 1979; Carter *et al.*, 1982). In addition, the numbers of sycamore aphids on trees in the Glasgow area and those caught in suction traps throughout Britain are correlated with one another, which indicates that good and bad years for this aphid are general throughout Britain. However, the further apart the traps the weaker the correlation, possibly reflecting regional differences in weather from year to year (Figure 8.18). Thus it is possible, as with the distribution of sycamore aphids between leaves on a tree, that the dramatically changing spatial and temporal mosaic of aerial population densities mainly just reflects a similar changing mosaic of habitat quality at ground level. This may not be true but only empirical studies can settle the issue.

Summarizing, the abundance of aphids, not surprisingly, is dramatically affected by changes in weather, a major disturbing factor. Aphid numbers are

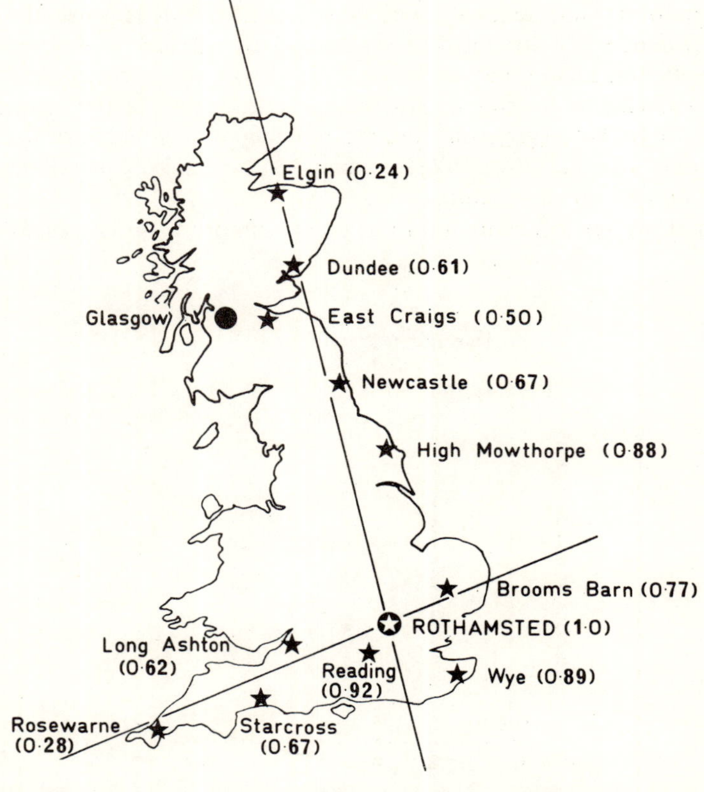

Figure 8.18 The location of some of the Rothamsted Insect Survey suction traps in the United Kingdom and the correlation coefficients between the number of sycamore aphids caught by these traps and the Rothamsted trap.

regulated by a hierarchy of different density-dependent processes controlling population growth at various levels of abundance. The dramatic spatial and temporal changes in aerial abundance over wide areas, so strikingly portrayed in the maps published by Taylor and his colleagues, are very suggestive of large-scale population movement. However, only if these changes in aerial abundance can be related to what is occurring on the host plants, and then subjected to experimental analysis, can we determine the extent to which movement shapes the spatial dynamics of aphid populations.

9 Community structure and species diversity

Although there are many studies of aphids at the individual and population levels there have been few attempts to study interactions between aphid and plant populations or even between aphid species, especially in the field. This is partly due to the difficulty of identifying aphids, especially the highly polymorphic species, and the tendency to regard plants solely as a source of food. However, as keys for the identification of aphids become available and the role of plants in aphid biology is more fully appreciated (Dixon, 1977), there is likely to be an increasing interest in this aspect of aphid ecology.

Mutualism between plants and aphids

Large quantities of honeydew are produced by aphids, which in some species contains a high proportion of the trisaccharide melezitose. This led Owen and Wiegert in a series of papers (Owen, 1978, 1980*a*; Owen and Wiegert, 1976, 1981) to propose that plants benefit from being eaten, and in particular, plants release surplus sugars by enlisting the help of aphids. The sugar in the form of honeydew is utilized by free-living nitrogen fixing bacteria in the soil, which increase in number beneath aphid-infested plants and make more nitrogen available to these plants. Melezitose, or a particular mixture of sugars in honeydew, are thought to have an optimal affect on nitrogen fixation. Thus these authors view aphids as a necessary 'part' of a plant, releasing surplus sugars that promote a better supply of nitrogen.

The addition of the four sugars commonly found in honeydew—fructose, glucose, melezitose and sucrose—to soil at rates equivalent to those recorded beneath trees (Llewellyn, 1972) causes an increase in the abundance of microbial populations, at least in woodland soils (Dighton, 1978*a*, *b*), and fructose is more effective at promoting nitrogen fixation than melezitose (Petelle, 1980). However, as single sugars rather than a mixture were used Petelle's results do not refute Owen's hypothesis (Owen, 1980*b*). Nevertheless, whatever the positive effect, if any, of honeydew in promoting the availability of nitrogen for plant growth one cannot ignore the adverse effect aphids have on plant growth and seeding.

Aphids and plant growth

As relatively few aphids gall or distort leaves their presence often goes unnoticed. However, they can become very abundant (p. 102), even on trees; for example, 116 000 leaves of a 20 m sycamore tree may be infested with 2.25×10^6 aphids, equivalent in mass to a large rabbit. It is not surprising, therefore, that infested trees achieve less growth than uninfested trees.

Trees vary in their response to aphid infestation. Although aphids do not affect the number of leaves borne by lime, oak or sycamore, sycamore produces smaller leaves, which contain more nitrogen, when heavily infested in spring (Figure 9.1). The width of the annual rings of sycamore is positively correlated with the average size of the leaves, and negatively, with the numbers of aphids on the tree throughout a year (Figure 9.2). In the absence of aphids some sycamore trees could produce as much as 280% more stem wood (Dixon, 1971*e*). Lime and oak aphids hatch later, relative to the time of bud burst of their host trees, than the sycamore aphid does and, as a consequence, rarely become abundant enough before the leaves are fully grown to affect their growth. Similarly, they do not affect the above-ground growth in girth and length. However, infested saplings of lime and oak often weigh less at the end of a year than they did at the start, mainly due to a reduction in the mass of their roots (Figure 9.3) (Dixon, 1971*f*).

Aphid infestation causes early leaf fall in all three species and, in oak and sycamore, results in the leaves becoming a darker green. In oak this is a consequence of a 25% increase in the quantity of both chlorophylls A and B. Associated with this is an increased dry matter production, which in sycamore can be 1.7 times greater in infested than uninfested leaves. Following years of heavy aphid infestation lime and sycamore break their buds later than usual,

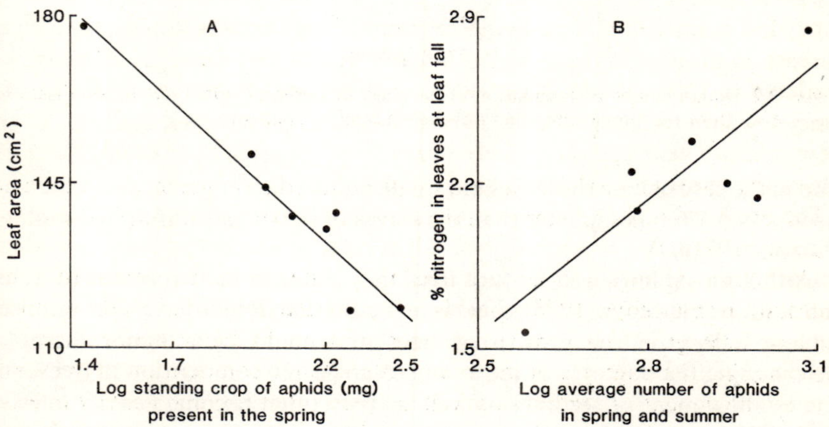

Figure 9.1 The average area of sycamore leaves in relation to the standing crop of sycamore aphids on the leaves in spring (*A*), and the percentage of nitrogen in the leaves at leaf fall in relation to the average number of sycamore aphids on the leaves in spring and summer (*B*).

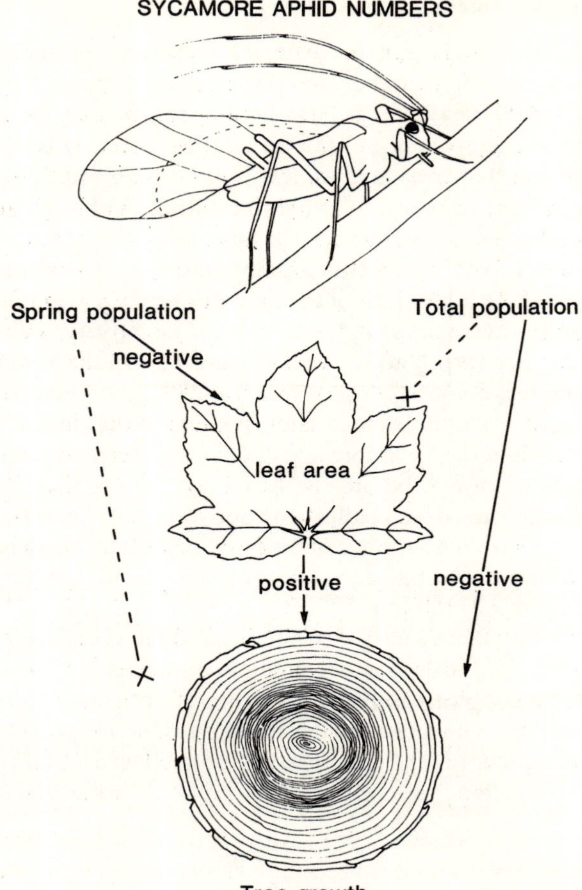

Figure 9.2 Diagrammatic representation of the effect of sycamore aphid numbers on leaf area and the width of the annual rings laid down each year by sycamore.

and in the case of lime the leaves are smaller and a darker green, and have a net production 1.6 times greater than the leaves of previously uninfested saplings (Dixon, 1971*e*,*f*).

Although saplings and mature trees may differ in their reaction to aphid infestation (Llewellyn, 1975), there is no doubt that aphids have a pronounced adverse effect on the growth of trees and could be a major factor in determining the outcome of intra- and interspecific competition in trees, and the establishment of seedings as well, as they often become heavily infested with aphids that fall from the parent trees. Some trees can compensate in part at least for the nutrient drain imposed by aphids and it remains to be shown that aphids affect the genetic fitness of mature deciduous trees. However, high

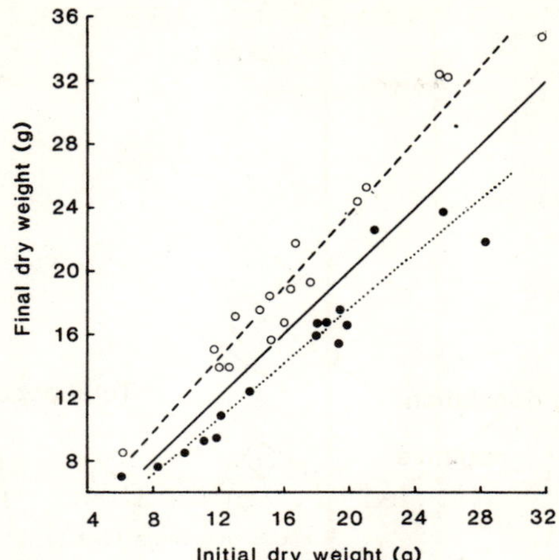

Figure 9.3 The effect of aphid infestation on the growth of oak saplings: their dry weight in autumn in relation to their estimated dry weight in the previous spring for saplings infested (●) and kept free of aphids (○) (—— no change in weight).

numbers of the spruce aphid (*Elatobium abietinum*) can cause the defoliation of spruce, which, if recurrent, can cause the death of the tree.

Aphids and the seeding of plants

The number and quality of seeds produced by plants have been used as a measure of genetic fitness. As many trees do not fruit and seed equally every year but at irregular intervals, they are not convenient for studying the effect of aphids on seed production. However, annual, biennial and short-lived perennial plants are more convenient for measuring both the total seed production and any beneficial effect of the proposed honeydew-increased nitrogen fixation.

Aphids affect the yield and quality of the seed of several crop plants. For example, *Aphis fabae* reduces the number of seeds produced by bean plants (*Vicia faba*) by up to 86% and the average weight of a bean by 45% (Banks and Macaulay, 1967). Similarly, aphids reduce the yield of cereals (Figure 9.4) (Vereijken, 1979).

Thus aphids do affect the reproductive potential of crop plants, even though, as in the case of cereals, they are given high levels of nitrogen fertilizer. As in a nutrient-rich environment aphid- infested plants still grow more slowly than uninfested plants, they are likely to suffer even more markedly from aphid infestation in nutrient-limited environments (Figure 9.5).

As well as imposing a severe nutrient drain on plants, aphids transmit viruses, which can also adversely affect the fitness of plants. Therefore, there is no evidence to support the contention that plants infested with aphids are

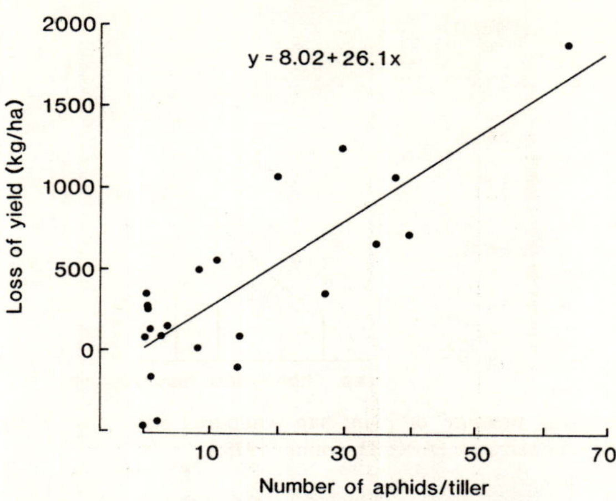

Figure 9.4 Loss of yield of wheat in relation to the peak number of *Sitobion avenae* per tiller. (After Vereijken, 1979).

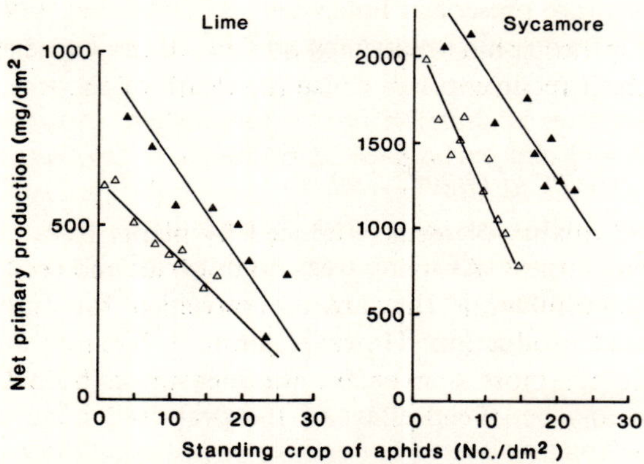

Figure 9.5 The net annual primary production per unit area of leaf in relation to the average number of aphids on the leaves during the year for lime and sycamore saplings watered with either full- (▲) or a half-strength (△) nutrient solutions. (After White, 1970.)

fitter than those that are free of aphids. If melezitose does not improve plant fitness overall, what is its function?

Melezitose

The difference in the osmotic concentration of the haemolymph of aphids and the phloem sap they ingest could impose a severe osmotic stress. However, aphids can substantially reduce the osmotic concentration of the ingested fluid, as honeydew is similar in osmotic concentration to the haemolymph

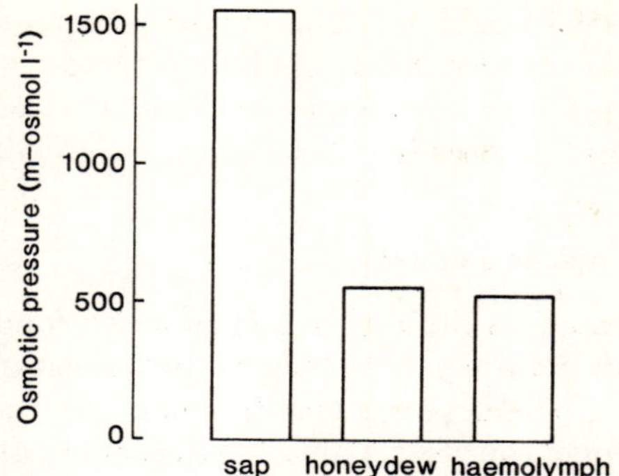

Figure 9.6 The osmotic pressure of plant sap compared with that of the honeydew and haemolymph of *Myzus persicae*. (After Downing, 1978.)

(Downing, 1978) (Figure 9.6). A factor in this is the conversion of mono- and disaccharides into trisaccharides like melezitose, which reduces the number of molecules in solution and thus its osmotic concentration. However, the quantity of melezitose present in honeydew, 45% (Michel, 1942; Bacon and Dickinson, 1957), is insufficient to account for all of the drop in osmotic concentration.

Another suggestion is that the ability to produce melezitose and other trisaccharides, which are particularly attractive to ants (Duckett, 1974), has been selected for because ant attendance has advantages for aphids (Kiss, 1981). A serious flaw in this hypothesis is that melezitose is present in the honeydew of several aphids, for example, the lime aphid (*Eucallipterus tiliae*) that are not ant-attended.

Energy and nutrient flow

A large amount of energy is used annually by aphid populations. Lime aphids consume 1.5×10^7 J m^{-2} yr^{-1} (Llewellyn, 1972), compared with 3×10^6 J m^{-2} yr^{-1} for grazing bullocks (Macfadyen, 1964) and 6.4×10^5 J m^{-2} yr^{-1} for browsing oak tree caterpillars (Varley, 1967). Thus aphids may extract much more energy per unit area than grazers and browsers, and do so without consuming any of the plant structure. However, of the food energy ingested by aphids little is used for growth and as much as 90% is excreted. Therefore, although relatively little of the food energy ingested is available to the natural enemies of aphids, a large proportion passes to the decomposers. The energy content of the honeydew falling from a lime tree in the course of a year is equivalent to that in the leaves at leaf fall (Llewellyn, 1970). In addition the percentage nitrogen content of the leaves at leaf fall is much higher (cf. Figure 9.1*B*) in years when aphids are abundant. Thus, fluctuations in aphid numbers have a dramatic effect on the quantity and quality of the food available to the decomposers. Although relatively little of the total energy ingested by aphid populations is available to their natural enemies, nevertheless aphids can become so abundant that they can give rise to 'plagues' of natural enemies (p. 102).

Mutualism between aphids and ants

The mutualism is based on the ants obtaining a rich food supply from the aphids and the aphids protection from their natural enemies. Not all aphids are ant-attended, nor do all ants attend aphids, and myrmecophilous species of aphids differ in their dependence on ants. Myrmecophilous species of aphids that live above ground appear to be adapted to ant attendance as they are often conspicuously coloured, gregarious and do not avoid an approaching ant, predator or parasite (Dixon, 1958) and myrmecophilous species generally have poorly-developed structural adaptations for defence against natural enemies (Way, 1963). The advantage of ant attendance for aphids is most striking for the larger species of aphids, which are particularly vulnerable to natural enemies while withdrawing their long stylets (p. 28). Not surprisingly, therefore, most large species of aphids are ant-attended and some like *Stomaphis quercus* are found only in association with certain species of ants.

Ants can have pronounced effects on the abundance of aphids. For example, numbers of the ant-attended *Periphyllus testudinaceus* on sycamore are higher on branches foraged by the ant *Formica rufa* than on adjacent branches from which they are excluded. However, the other species of aphid on sycamore, *D. platanoidis*, which is not ant-attended, does better in the absence of ants (Figure 9.7) (Skinner and Whittaker, 1981). Similarly, of the four aphids on

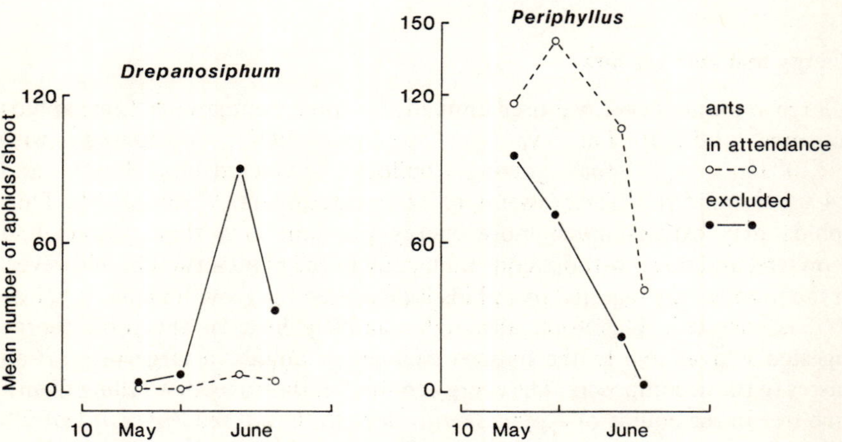

Figure 9.7 Mean numbers of *Drepanosiphum platanoidis* and *Periphyllus testudinaceus* on shoots of sycamore that are ant-attended and from which ants are excluded. (After Skinner and Whittaker, 1981.)

fireweed three do better, and one worse, when on plants that are ant-attended (Addicott, 1979).

When ant-attended, *A. fabae* increases its uptake of phloem sap and so increases its production of honeydew by up to 50%. The greater growth and fecundity rates of ant-attended *A. fabae* are attributed to the improved nutrition resulting from increased feeding (El-Ziady, 1960) and to their tendency to disperse less and, consequently, for most of them to aggregate and feed on the youngest parts of their host plant (Banks, 1957; Banks and Nixon, 1958; Banks and Macaulay, 1967). It is possible that in the absence of ants such aphids use more of their resources for dispersal to increase the chance of survival of their clones. The higher proportion that develop into alatae in the absence of ants (Banks, 1958; El-Ziady, 1960) also lends support to the idea of an inborn trade-off between dispersal and reproduction, with reproduction favoured when the possibility of being killed by natural enemies is reduced. Alternatively it may be due to the 'tranquillizing' action of the ants reducing the restlessness of the aphid and thus tactile stimulation necessary for alate determination (p. 38) (Way, 1963).

Newly-founded colonies of *Aphis varians* and *A. helianthi* on fireweed, and of *Pterocomma populifoliae* on aspen, are more likely to persist if they are found and tended by ants within the first week. Exclusion of ants causes such colonies to go extinct faster than ant-attended colonies. This is due to the ants attacking and killing aphid predators like coccinellid beetles (Sanders and Knight, 1968; Addicott, 1979). Similarly the ant *Lasius niger* affords small colonies of *A. fabae* protection against a range of natural enemies (Banks, 1962).

Thus the dynamics of colonies of ant-attended species of aphid are very dependent on the availablity of ants. Unattended small colonies are more likely to decline than attended colonies. However, in large colonies, the effect is reversed, with ant-attended colonies more likely to decline than unattended colonies. Since the magnitude of the effect of ants on aphids is large at densities or times when a colony is most likely to go extinct, and since it is smaller when extinction is less likely, this acts as a stabilizing factor (Addicott, 1979). However, as ant attendance can result in a rapid increase in the size of the aphid colonies, as well as a 50% increase in the rate of feeding of the individual aphids, the combined effect could be a more rapid deterioration in the quality of the host plant for the aphid. It seems likely that it is this, rather than the presence of ants, that results in the decline of the larger colonies. In those species of aphids that are only found in association with ants it is possible that the ants keep the aphids at densities below that which is harmful to their host plants and may even keep the total numbers at levels that satisfy the food requirements of the ant colony (Way, 1963).

In foraging for honeydew, ants are likely to be attracted to the richer sources, only exploiting poor sources when honeydew is in short supply. Thus there could be competition between aphids for ants. There is some support for

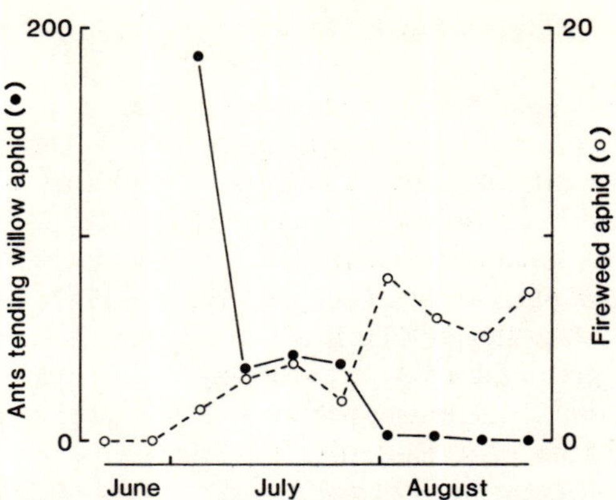

Figure 9.8 Numbers of ants attending colonies of the willow aphid, *Aphis farinosa*, and the fireweed aphid, *Aphis varians*, in relation to time of year. (After Addicott, 1978*b*.)

this in that colonies are more likely to go extinct if they are close to plants occupied by ant-attended aphids. In addition, there is some evidence that ants switch their attention to other species of aphids as a source of honeydew when the numbers of the favoured species of aphid decline (Figure 9.8). However, further work is required to provide unequivocal evidence that ants forage optimally.

Competition

Aphid species may feed on different parts of a plant at different times of the year (Figure 9.9). The spatial separation of species may enable them to occur at the same time on the same plant, but does not prevent the aphid species from competing. In the absence of a species its niche is likely to be occupied by another species of aphid, providing its morphology or behaviour does not constrain it. *A. fabae*, living on bean, changes its feeding site from the stem to the leaves, and from the upper to the lower surface of leaves, when *Acyrthosiphon pisum* is present (Salyk and Sullivan, 1982). However, in other coexisting species there is no indication that the presence of one species alters the feeding position of the other (Addicott, 1978*a*).

Although the feeding positions on a plant may not overlap, nevertheless the aphids are feeding on a common resource—phloem sap. Intraspecific competition for this resource can be intense and result in longer developmental times, smaller aphids or a switch to the development of another morph better adapted to dispersal. In this way, *Macrosiphum valerianae* adversely affects the weight of *Aphis varians* when both are feeding on the same plant (Addicott, 1978*a*); and in the presence of the sycamore aphid, *D. platanoidis*, proportionately more of the other aphid on sycamore, *Periphyllus testudinaceus*, develop into alatae (Shearer, 1976). The in-

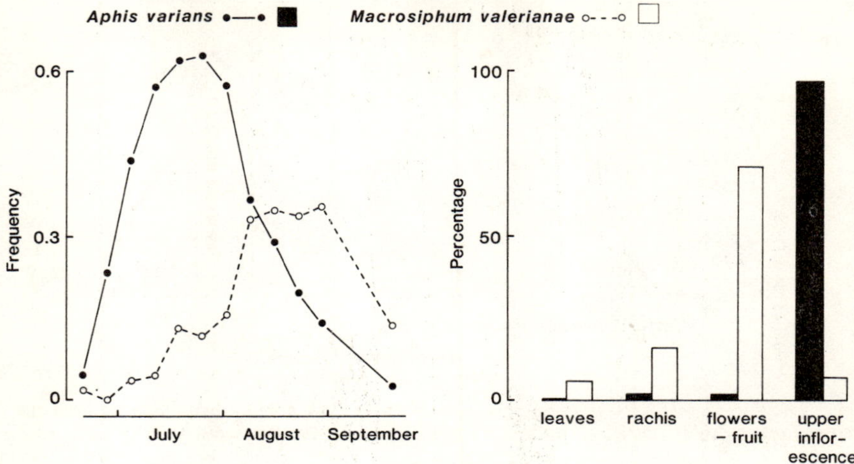

Figure 9.9 Frequency of infestation of fireweed plants by *Aphis varians* and *Maerosiphum valerianae* in relation to time of year, and the distribution of each aphid on the plants.

terspecific effect on alate production is less, however, than the intraspecific effect.

The rain of honeydew from *Chromaphis juglandicola* that feeds on the undersides of walnut leaves is thought to prevent the colonization of the upper surface of the leaves by *Calaphis juglandis* (Stroyan, 1977); this is a form of interference. However, as *C. juglandis* prefers the terminal leaves and the ants that attend it are likely to reduce the numbers of *C. juglandicola* in the immediate vicinity, the two species are likely to coexist.

On the other hand, aphids do not always adversely affect one another. In autumn, *Periphyllus acericola* induce rapid local senescence of the sycamore leaves on which they feed, and individuals of *D. platanoidis* feeding alongside them are larger than those that feed on adjacent leaves without *P. acericola.*

The cotton aphid, *Aphis gossypii*, and the chrysanthemum aphid, *Macrosiphoniella sanborni*, both occur on chrysanthemums in greenhouses. They feed on different parts of the plant, *M. sanborni* on the growing stem, and *A. gossypii* on the young leaves. This habitat segregation is less marked, however, when large numbers of aphids cause the plant to stop growing, and no young growth is available. The presence of one species inhibits the growth and population density achieved by the other species (Figure 9.10). Also, when in a mixed population *A. gossypii* only produces 55% and *M. sanborni* 78% of the number of alatae they each produced when not competing (Figure 9.11). It is also interesting to note that the equilibrium populations achieved in the mixed and single-species populations are similar. This suggests that both species affect the plant similarly and do not affect the plant's carrying capacity.

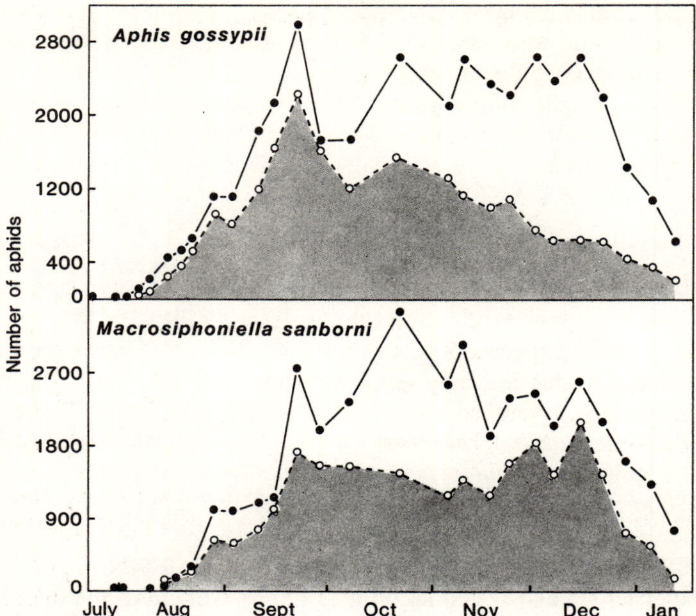

Figure 9.10 The numbers of *Aphis gossypii* and *Macrosiphoniella sanborni* on chrysanthemums in single species (●——●) and mixed species populations (○—— ○). (After Tamaki and Allen, 1969.)

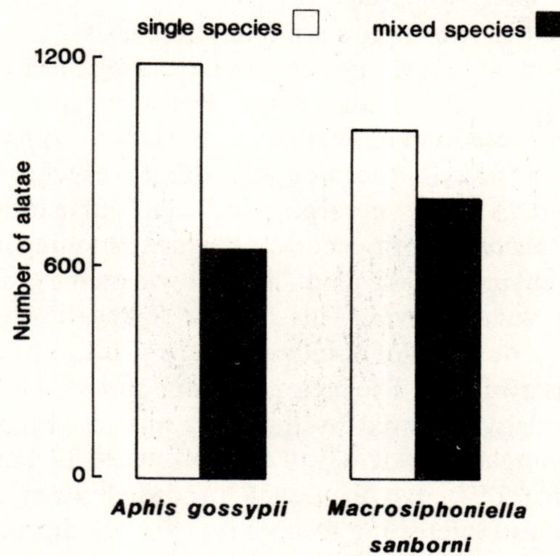

Figure 9.11 The number of alatae produced by *Aphis gossypii* and *Macrosiphoniella sanborni* in single (□) and mixed species (■) populations on chrysanthemums. (After Tamaki and Allen, 1969.)

Although aphids living on a particular species of plant often show marked habitat segregation and associated behavioural and morphological differences, in feeding on phloem sap they are competing for a limited resource. Although interspecific competition is unlikely to be as marked as intraspecific competition it is nevertheless likely to be an important process in the dynamics of some aphid species.

Species diversity and plant architectural complexity

Generally, the more structurally complex plants, like trees, are regularly infested with more species of aphids than herbaceous plants. Although grasses can have one aphid species on the flower and different species on the stem, leaves and roots, this is rarely achieved and then only for short periods. However, on trees, several species of autoecious aphids regularly coexist, e.g. *Tuberculoides annulatus, Thelaxes dryophila, Lachnus roboris* and *Stomaphis quercus* on oak; and *Betulaphis quadrituberculata, Euceraphis punctipennis, Monaphis antennata* and *Symydobius oblongus* on birch.

The increase in structural complexity that results from increase in the size of an individual plant can sometimes result in the loss of species. *Cinara todocola* tends to infest young todo-fir, and is rarely found on plants more than 14 years old. This is thought to be because the aphids feed at the tips of the branches and so, on the taller trees they are more exposed to extremes of temperature and less well protected from their natural enemies by ants (Figure 9.12) (Yamaguchi, 1976). Thus in addition to having more species of aphids overall, trees are likely to both gain and lose species as they increase in size and structural complexity.

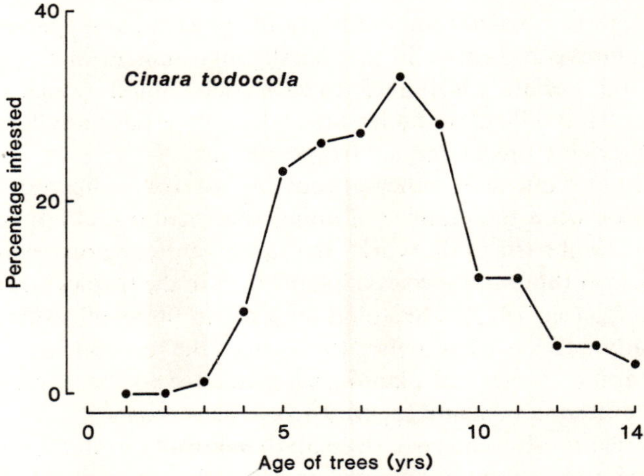

Figure 9.12 The percentage of todo-fir plants infested with *Cinara todocola* in relation to their age. (After Yamaguchi, 1976.)

Why are there so few species of aphids, especially in the tropics?

Most aphids are monophagous, have similar sensory and dispersal powers and are not attracted to their host plants from a distance. Host finding is a random process. Thus the concept of plant apparency is useful when considering why relatively few plants have aphids, and also the somewhat surprising paucity of aphids in the tropics. A plant that is plentiful for long periods of time, and over a large area, is apparent, and is more likely to have aphids than less apparent species of plants.

About 4000 species of aphids have so far been described and most of these come from the temperate regions of the world. In the tropics and subtropics Aleyrodidae and Coccoidea seem to replace the aphids. The respective distributions of all three groups are assumed to reflect their ability to survive in the physical conditions that prevail in the temperate and tropical regions (Bodenheimer and Swirski, 1975). However, aphids associated with crops have often retained their pest status when introduced from the temperate regions into tropical and subtropical regions of the world. There are also holocyclic endemic species in the tropics and subtropics, well adapted to the climatic conditions that prevail there. For example, in Australia species of *Schoutedenia* (Greenideinae), *Sensoriaphis* and *Neophyllaphis* (Drepanosiphinae) avoid the rigours of the summer by producing eggs in spring (Hales, 1976). In contrast the anholocyclic *Cervaphis schouteniae* (Greenideinae) in the vicinity of Calcutta, India, achieves its highest rates of increase during the hottest (30° C +) and driest period of the year (Agarwala and Dixon, 1985). Indeed species from all the aphid subfamilies are found in the tropics and subtropics. Although continuous parthenogenesis is commoner in species living in the tropics and subtropics than in the temperate regions, nevertheless sexual forms have been recorded for 20% of the Indian aphids (Agarwala, personal communication). Thus as some species of aphids are able to survive and thrive in the 'harsh' environment of the tropics and subtropics and speciation is unlikely to have been limited by a lack of genetic recombination it is difficult to understand why aphids have not flourished as a group on the rich tropical and subtropical floras.

Aphids differ from most other groups of insects by showing an inverse relationship between the number of aphid species and the number of plant species in different parts of the world. In the temperate regions there are more aphid species per thousand species of plants than in the tropics and subtropics (Table 12 in Eastop, 1973). The aphid faunas and floras of several countries are sufficiently well known to define the form of the relationship between the number of aphid species per plant species, relative to the number of plant species per unit area of land (Figure 9.13).

A similar relationship can be derived mathematically from the observations that aphids are mostly monophagous and find their host plants by random

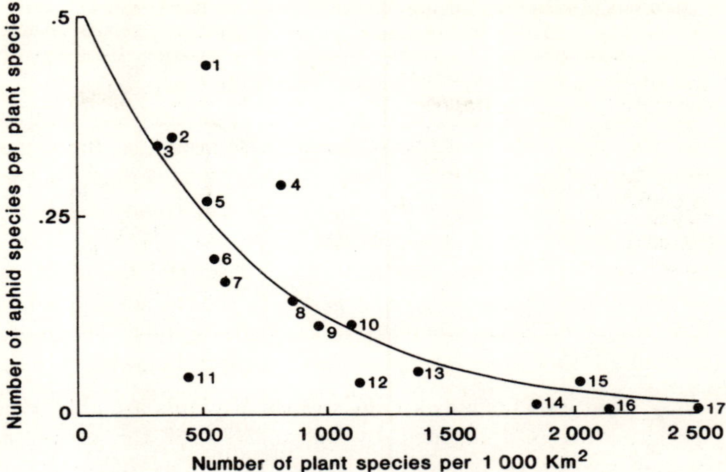

Figure 9.13 The number of aphid species per plant species in relation to the number of plant species per unit area of land surface for seventeen countries. 1, The Netherlands; 2, United Kingdom; 3, Nordic countries; 4, Czechoslovakia; 5, Poland; 6, Azores; 7, Russia (European); 8, Romania; 9, Portugal; 10, Madeira; 11, New Zealand; 12, Turkey; 13, Spain; 14, Cuba; 15, India; 16, Australia; 17, South Africa.

search (Dixon *et al.*, 1985):

$$Sa/Sp \simeq Ke^{-C_{crit}Sp}$$

where *Sa* is the number of aphid species, *Sp* the number of plant species, *K* the average number of aphid species per plant species whose cover is equal to or greater than C_{crit}, which is the proportional ground cover a plant species must achieve if it is to support an aphid species. Thus in the tropics and subtropics, where the floral diversity is very high, few species of plants are apparent enough to sustain an aphid species. In this context it is interesting to note also that proportionately more aphids are polyphagous in the tropics than in the temperate regions (Holman, 1971). By living on a number of uncommon plant species, whose collective apparency is above the threshold necessary to sustain an aphid species, polyphagous aphids possibly compensate for the 'rarity' of their individual host-plant species. In the temperate regions there are fewer plant species than in the tropics, and the commoner ones are apparent enough to sustain one or more species of aphids. This, combined with the greater niche diversity produced by seasonal changes in habitat quality in temperate regions (p. 78), possibly accounts for the greater number of species of aphids that have evolved in temperate regions.

Consideration of the world-wide distribution of aphids not only provides an answer for the paucity of aphids in the tropics and sub-tropics, in terms of

Table 9.1 Percentage of common, scattered and rare plants of three plant families, that are host to monophagous species of aphids in Czechoslovakia

Plant family	*Number of species*	*Abundance*		
		Common	Scattered	Rare
Compositae	226	71	31	14
Ranunculaceae	72	33	30	8
Umbelliferae	78	65	43	13
		Average percentage		
		65	33	12

aphid biology, but also for why so few species of plants are able to sustain aphid species. This is confirmed by an analysis of the occurrence of monophagous species of aphids on species of plants belonging to the Compositae, Ranunculaceae and Umbelliferae in Czechoslovakia, where the aphid fauna and flora are well known (Table 9.1). Proportionately more monophagous species of aphids are associated with common than with scattered or rare species of plants. The presence of aphids on the rare species of plants does not invalidate the idea because many of these plants are common in adjacent countries (Dixon *et al.*, 1985).

Polyphagy

Most aphids are behaviourally, morphologically and physiologically adapted to living on a particular species or genus of plants. This high degree of monophagy is very characteristic of aphids. Accepting that monophagy is a successful way of life for aphids, it remains to account for the existence and success of the relatively few species that are polyphagous, and their ability to coexist with monophagous species.

In aphids, polyphagy is confined to a few species belonging to one subfamily, the Aphidinae, but, what is possibly more important, it is confined to aphids living on herbaceous plants. Changes in weather result in changes in the relative abundance of herbaceous plants from year to year, and from place to place, within a plant community (Grubb *et al.*, 1982). The relative abundance of trees is less likely to be affected by year to year changes in weather. Therefore aphids living on herbaceous plants are likely to experience years in which their host plants are rare or of poor quality, or both. In such years the specific aphids associated with the plants will decline in numbers, and are likely to take some time to recover when their host plants become more abundant again. Aphids living on trees, although affected by changes in the weather, nevertheless tend to be relatively abundant, at least for part of any year. Thus in communities of herbaceous plants, changes in the relative abundance of the species from year to year possibly prevent some monophagous species of aphids from fully exploiting their host plants every year. Some species of plants may become, or are, so uncommon in certain years that their aphids become extinct or they do not have their own specific aphids. However, a polyphagous species of aphid may be able to survive by exploiting a number of these plants concurrently because their total abundance will be high and they will often not be fully exploited by monophagous species of aphids.

To date, unfortunately, there are no long-term population studies of monophagous species of aphids living on herbaceous plants. Therefore, the assumption of greater population stability and fuller exploitation of resources on trees, relative to herbaceous plants, is untested.

Summarizing, there is no evidence for the suggested mutualism between aphids and plants. In fact aphid infestation adversely affects the fitness of many plants. The mutualism between ants and some aphid species, however, is well established. Most of the large amount of energy-providing substances in the food ingested by aphids is excreted as honeydew, which mainly passes directly to the decomposers. Although relatively little of the energy taken by aphids is used for growth, nevertheless they can become very abundant and so support large numbers of natural enemies. Interspecific competition between aphids has been recorded under experimental conditions and is likely to be an important factor in the field. More species of aphids are associated with individual species of trees than herbaceous plants, which may be a consequence of the greater architectural complexity of trees. The greater number of species of aphids in the temperate regions can possibly be accounted for mainly in terms of floral diversity and aphid biology. In the tropics the high floral diversity means that relatively few species of plants exceed the critical level of cover necessary to sustain an aphid species. Polyphagy has possibly enabled a few species of aphids to exploit a number of herbaceous plants that vary in abundance and availability for exploitation between years, but in combination exceed the critical level of cover necessary to support a species of aphid. That aphids can be characterized as opportunistic species indicates that aphids respond less to the permanence of their host plants than to their dramatically changing quality which is true for aphids living on both herbaceous and woody plants.

10 Epilogue

Parthenogenesis and feeding on phloem sap, which both developed early in the evolution of aphids, have been major factors shaping the ecology of the group. Parthenogenesis led to the evolution of clonal polymorphism and the telescoping of generations. The latter, along with parthenogenesis, resulted in a prodigious rate of increase that is of great selective advantage when colonizing the temporarily unexploited and highly favourable habitats available to all species of aphids at certain times of the year. However, the simultaneous commitment to growth and reproduction puts a severe constraint on the length of time for which aphids can survive without feeding, and thus the time they can spend searching for host plants. Small size and host specificity, both adaptations to feeding on phloem sap, also impose constraints. As small insects have little control over their direction of flight, there is no selective advantage in being able to recognize a potential host plant at a distance. The short period for which aphids can survive without food, their high degree of host specificity, and the random location of host plants have limited the number of species of aphids and confined them to areas where plant diversity is relatively low, i.e. the temperate regions. However, because aphids are opportunists adept at rapidly increasing in numbers when conditions are favourable, individual species can and often do become very numerous.

Their high degree of host specificity and close association with their host plants leads one to expect that coevolution is a major force generating patterns in aphid ecology. Although there is abundant evidence of how aphids adapt to their host plants there is little unequivocal evidence of how plants have adapted to aphids, possibly because plants are subject to attack by a very wide range of organisms, and selection has favoured the development of plant defences effective against generalist phytophages rather than the specialists.

Thus it is only by understanding the nature of the constraints associated with the evolution of certain features of a group that we will come to understand its ecology. The complexity of the environment in which aphids live often obscures and distracts us from the major constraints that have shaped their ecology and evolution, but makes the story of aphids an apparently everlastingly entertaining one to unfold.

Appendix: Scientific names of aphids referred to in the text

Acyrthosiphon kondoi Shinji
Acyrthosiphon pisum (Harris)—Pea aphid
Acyrthosiphon pisum subsp. *destructor* Johnson
Amphorophora rubi subsp. *idaei* Borner—Raspberry aphid
Aulacorthum circumflexum (Buckton)—Lily aphid
Aphis craccivora Koch
Aphis cytisorum Hartig—Black broom aphid
Aphis fabae Scopoli—Bean aphid
Aphis farinosa Gmelin—Small willow-aphid
Aphis gossypii Glover—Cotton aphid
Aphis helianthi Monell
Aphis nerii Boyer de Fonscolombe—Oleander aphid
Aphis pomi de Geer—Apple aphid
Aphis rumicis Linnaeus—Dock aphid
Aphis sambuci Linnaeus—Elder aphid
Aphis urticata Gmelin (≡ *Pergandeida stanilandi* Laing)
Aphis varians Patch
Astegopteryx styracicola Takahashi
Betulaphis quadrituberculata (Kaltenbach)
Brevicoryne brassicae (Linnaeus)—Cabbage aphid
Callaphis juglandis (Goeze)—Large walnut aphid
Callipterinella minutissima (Stroyan)
Cavariella aegopodii (Scopoli)—Willow-carrot aphid
Cervaphis schouteniae van der Goot
Chromaphis juglandicola (Kaltenbach)—Walnut aphid
Cinara pilicornis (Hartig)—Spruce shoot aphid
Cinara todocola (Inouye)—Todo fir aphid
Cryptomyzus galeopsidis (Kaltenbach)—Currant aphid
Drepanosiphum acerinum (Walker)
Drepanosiphum dixoni Hille Ris Lambers
Drepanosiphum platanoidis (Schrank)—Sycamore aphid
Dysaphis devecta (Walker)—Rosy leaf-curling aphid
Dysaphis plantaginea (Passerini)—Rosy apple-aphid
Elatobium abietinum (Walker)—Green spruce aphid
Epipemphigus niisimae (Matsumura)
Eriosoma yangi Takahashi
Eucallipterus tiliae (Linnaeus)—Lime aphid
Euceraphis punctipennis (Zetterstedt)—Birch aphid
Lachnus iliciphilus (del Guercio)
Lachnus roboris (Linnaeus)
Lipaphis erysimi (Kaltenbach)
Macrosiphoniella sanborni (Gillette)
Macrosiphum euphorbiae (Thomas)—Potato aphid
Macrosiphum rosae (Linnaeus)—Rose aphid
Macrosiphum valerianae (Clarke)
Masonaphis maxima (Mason)—Thimbleberry aphid
Megoura viciae Buckton—Vetch aphid
Metopolophium dirhodum (Walker)—Rose grain aphid
Microlophium carnosum (Buckton)—Nettle aphid
Monaphis antennata (Kaltenbach)
Myzus ligustri (Mosley)—Privet aphid
Myzus persicae (Sulzer)—Green peach aphid
Nasonovia ribisnigri (Mosley)—Lettuce aphid
Patchiella reaumuri (Kaltenbach)
Pemphigus betae Doane—Sugar beet root aphid
Pemphigus bursarius (Linnaeus)—Lettuce root aphid
Pemphigus dorocola Masumura
Pemphigus treherni Foster

Periphyllus acericola (Walker)
Periphyllus testudinaceus (Fernie)—European maple aphid
Phorodon humuli (Schrank)—Damson Hop aphid
Pseudoregma alexanderi (Takahashi)
Pterocomma populifoliae (Fitch)
Rhopalosiphum insertum (Walker)—Apple-grain aphid
Rhopalosiphum maidis (Fitch)—Corn leaf aphid
Rhopalosiphum padi (Linnaeus)—Bird cherry-oat aphid
Schizaphis graminum (Rondani)—Greenbug
Sitobion avenae (Fabricius)—Grain aphid
Stomaphis quercus (Linnaeus)
Symydobius oblongus (von Heyden)
Thecabius affinis (Kaltenbach)—Poplar buttercup aphid
Thelaxes dryophila (Schrank)
Therioaphis trifolii (Monell)—Alfalfa aphid
Triassoaphis cubitus Evans
Tuberculoides annulatus (Hartig)—Oak aphid
Tuberolachnus salignus (Gmelin)—Large willow aphid

References

Addicott, J. F. (1978*a*) Niche relationships among species of aphids feeding on fireweed. *Can. J. Zool.* **56**:1837–1841.

Addicott, J. F. (1978*b*) Competition for mutualists: aphids and ants. *Can. J. Zool.* **56**:2093–2096.

Addicott, J. F. (1979) A multispecies aphid-ant association: density dependence and species-specific effects. *Can. J. Zool.* **57**:558–569.

Agarwala, B. K. and Dixon, A. F. G. (1985) Population trends of *Cervaphis schouteniae* van der Goot on *Microcus paniculata* and its relevance to the paucity of aphid species in India. *Proc. Ind. Acad. Sci.* (in press).

Akimoto, S. (1981) Gall formation by *Eriosoma* fundatrices and gall parasitism in *Eriosoma yangi* (Homoptera, Pemphigidae), *Kontyû* **49**:426–436.

Akimoto, S. (1983) A revision of the genus *Eriosoma* and its allied genera in Japan (Homoptera: Aphidoidea). *Insecta Matsumurana n.s.* **27**:37–106.

Anderson, R. M., Gordon, D. M., Crawley, M. J. and Hassell, M. P. (1982) Variability in the abundance of animal and plant species. *Nature* **296**:245–248.

Aoki, S. (1975) Descriptions of the Japanese species of *Pemphigus* and its allied genera (Homoptera: Aphidoidea). *Insecta Matsumurana n.s.* **5**:1–63.

Aoki, S. (1977) *Colophina clematis* (Homoptera, Pemphigidae), an aphid species with 'soldiers'. *Kontyû* **45**:276–282.

Aoki, S. (1979) Further observations on *Astegopteryx styracicola* (Homoptera: Pemphigidae), an aphid species with soldiers biting man. *Kontyû* **47**:99–104.

Aoki, S. (1980) Occurrence of a simple labour in a gall aphid, *Pemphigus dorocola* (Homoptera, Pemphigidae). *Kontyû* **48**:71–73.

Aoki, S. and Makino, S. (1982) Gall usurpation and lethal fighting among fundatrices of the aphid *Epipemphigus niisimae* (Homoptera, Pemphigidae). *Kontyû* **50**:365–376.

Aoki, S., Yamane, S. and Kiuchi, M. (1977) On the biters of *Astegopteryx styracicola* (Homoptera, Aphidoidea). *Kontyû* **45**:563–570.

Argandoña, V. H., Corcuera, L. J., Niemeyer, H. M. and Campbell, B. C. (1983) Toxicity and feeding deterrency of hydroxamic acids from gramineae in synthetic diets against the green bug, *Schizaphis graminum*. *Entomol. exp. appl.* **34**:134–138.

Argandoña, V. H., Luza, J. G., Niemeyer, H. M. and Corcuera, L. J. (1980) Role of hydroxamic acids in the resistance of cereals to aphids. *Phytochemistry* **19**:1665–1668.

Bacon, J. S. D. and Dickinson, B. (1957) The origin of melezitose: a biochemical relationship between the lime tree (*Tilia* spp.) and an aphid (*Eucallipterus tiliae* L.). *Biochem. J.* **66**: 289–299.

Banks, C. J. (1955) An ecological study of Coccinellidae (Col.) associated with *Aphis fabae* Scop. on *Vicia faba*. *Bull. ent. Res.* **45**:561–587.

Banks, C. J. (1958) Effects of the ant, *Lasius niger* (L.), on the behaviour and reproduction of the black bean aphid, *Aphis fabae* Scop. *Bull. ent. Res.* **49**:701–714.

Banks, C. J. (1962) Effects of the ant *Lasius niger* (L.) on insects preying on small populations of *Aphis fabae* Scop. on bean plants. *Ann. appl. Biol.* **50**:669–679.

Banks, C. J. (1965) Aphid nutrition and reproduction. *Rept. Rothamsted Expt. Sta.* (1964) 299–309.

Banks, C. J. and Macaulay, E. D. M. (1967) Effects of *Aphis fabae* Scop. and its attendant ants and insect predators on yields of field beans (*Vicia faba* L.). *Ann. appl. Biol.* **60**:445–453.

Banks, C.J. and Nixon, H. L. (1958) Effects of the ant, *Lasius niger* L., on the feeding and excretion of the bean aphid, *Aphis fabae* Scop. *J. exp. Biol.* **35**:703–711.

Barlow, C. A. (1962) The influence of temperature on the growth of experimental populations of *Myzus persicae* (Sulzer) and *Macrosiphum euphorbiae* (Thomas) (Aphididae). *Can. J. Zool.* **40**:145–156.

Barlow, N. D. and Dixon, A. F. G. (1980) *Simulation of Lime Aphid Population Dynamics*. Pudoc, Wageningen, 165 pp.

Behrendt, K. (1966) 'Population dynamics of *Aphis fabae* Scop. and the influences of Coccinellidae', in *Ecology of Aphidophagous Insects*, ed. I. Hodek, Academia, Prague, pp. 259–262.

Behrendt, K. (1969) Über langjährige Massenwechselbeobachtung an der Schwarzen Bohnenblattlaus, *Aphis fabae* Scopoli (Homoptera Aphididae). *Ber. 10 Wanderversammlung Dtsch. Entomol.* (1965) 335–354.

Behrendt, K. (1971) Zur Bedeutung der Dispersion für die Abundanz-dynamik von *Aphis fabae* Scop. (Homoptera: Aphididae) auf Zuckerrübenbestanden. *Zool. Jahrb. Abt. Syst. Oekol. Geogr. Tiere* **98**:418–454.

Bejer-Petersen, B. (1962) Peak years and regulation of numbers in the aphid *Neomyzaphis abietina* Walker. *Oikos* **13**:155–168.

Bell, G. (1980) The costs of reproduction and their consequences. *Amer. Nat.* **116**:45–76.

Bevan, D. and Carter, C. I. (1980) Frost proofed aphids. *Antenna* **4**(1):6–8.

Bintcliffe, E. J. B. and Wratten, S. D. (1982) Antibiotic resistance in potato cultivars to the aphid *Myzus persicae*. *Ann. appl. Biol.* **100**:383–391.

Blackman, R. L. (1976) Biological approaches to the control of aphids. *Phil. Trans. R. Soc. Lond. B.* **274**:473–488.

Blackman, R. L. (1979) Stability and variation in aphid clonal lineages. *Biol. J. Linn. Soc.* **11**:259–277.

Blackman, R. L. (1980) 'Chromosomes and parthenogenesis in aphids', in *Insect Cytogenetics*, ed. R. L. Blackman, G. M. Hewitt and M. Ashburner, 10th Symposium Roy. Ent. Soc. London, Blackwell, Oxford, pp. 133–148.

Blackman, R. L. (1981) 'Species, sex and parthenogenesis in aphids', in *The Evolving Biosphere*, ed. P. L. Forey, Cambridge University Press, pp. 75–85.

Blackman, R. L. and Takada, H. (1975) A naturally occurring chromosomal translocation in *Myzus persicae* (Sulzer). *J. Ent. (A)* **50**:147–56.

Blakley, N. (1982) Biotic unpredictability and sexual reproduction: Do aphid genotype-host genotype interactions favor aphid sexuality? *Oecologia (Berl.)* **52**:396–399.

Bodenheimer, F. S. and Swirski, E. (1957) *The Aphidoidea of the Middle East*. Weizmann Science Press, Jerusalem.

Bonnemaison, L. (1951) Contribution à l'étude des facteurs provoquant l'apparition des formes ailées et sexuées chez les Aphidinae. *Ann. Epiphyt.* **2**:1–380.

Bonnemaison, L. (1956) Déterminisme de l'apparition des larves estivales de *Periphyllus* (Aphidinae). *C.R. hebd. Séanc. Acad. Sci. Paris* **243**:1166–1168.

Bonnet, Ch. (1745) *Traité d'Insectologie*. Paris.

Börner, C. (1952) Europae centralis Aphides. *Mitt. Thür. Bot. Ges. Beiheft* **3,** Weimar.

Briggs, J. B. (1965) The distribution, abundance, and genetic relationships of four strains of the rubus aphid (*Amphorophora rubi* (Kalt.)) in relation to raspberry breeding. *J. hort. Sci.* **40**:109–117.

Bromley, A. K., Dunn, J. A. and Anderson, M. (1979) Ultrastructure of the antennal sensilla of aphids. I. Coeloconic and placoid sensilla. *Cell Tissue Res.* **203**:427–442.

Bromley, A. K., Dunn, J. A. and Anderson, M. (1980) Ultrastructure of the antennal sensilla of aphids. II. Trichoid chordotonal and campaniform sensilla. *Cell Tissue Res.* **205**:493–511.

Bromley, A. K.and Anderson, M. (1982) An electrophysiological study of olfaction in the aphid, *Nasonovia ribis-nigri*. *Entomol. exp. appl.* **32**:101–110.

Brown, M. I. (1975) *Intra-specific mechanisms regulating the numbers of the lime aphid.* Ph.D. thesis, University of Glasgow.

Burns, M. (1971) *Flight in the vetch aphid,* Megoura viciae *Buckton*. Ph.D. thesis, University of Glasgow.

Calow, P. (1978) *Life Cycles*. Chapman & Hall, London.

Campbell, A., Frazer, B. D., Gilbert, N., Gutierrez, A. P. and Mackauer, M. (1974) Temperature requirements of some aphids and their parasites. *J. appl. Ecol.* **11**:431–438.

Campbell, A. and Mackauer, M. (1977) Reproduction and population growth of the pea aphid (Homoptera:Aphididae) under laboratory and field conditions. *Can. Ent.* **109**:277–284.

Campbell, B. C. and Nes, W. D. (1983) A reappraisal of sterol biosynthesis and metabolism in aphids. *J. Insect Physiol.* **29**:149–156.

Carter, C. I. (1972) Winter temperatures and survival of the green spruce aphid. *For. Comm. For. Res.* **84**:1–10.

Carter, C. I. (1982) 'Susceptibility of *Tilia* species to the aphid *Eucallipterus tiliae*', in *Proc. 5th Int. Symp. Insect-Plant Relationships, Wageningen*, pp. 421–423.

Carter, N., Dixon, A. F. G. and Rabbinge, R. (1982) *Cereal Aphid Populations: Biology, Simulation and Prediction.* Pudoc, Wageningen.

Chambers, R. J. (1979) *Simulation modelling of a sycamore aphid population.* Ph.D. thesis, University of East Anglia.

Chambers, R. J. and Sunderland, K. D. (1983) 'The abundance and effectiveness of natural enemies of cereal aphids on two farms in Southern England', in *Aphid Antagonists*, ed. R. Cavalloro, Balkema, Rotterdam, pp. 83–87.

Chambers, R. J., Sunderland, K. D., Stacey, D. L. and Wyatt, I. J. (1982) A survey of cereal aphids and their natural enemies in winter wheat in 1980, *Ann. appl. Biol.* **101**:175–178.

Chambers, R. J., Wellings, P. W. and Dixon, A. F. G. (1985) Sycamore aphid numbers and population density. II Some Processes (in press).

Chapman, R. F., Bernays, E. A. and Simpson, S. J. (1981) Attraction and repulsion of the aphid, *Cavariella aegopodii*, by plant odors. *J. chem. Ecol.* **7**:881–888.

Cognetti, G. (1961) Citogenetica della partenogenesi negli Afidi. *Archo. zool. ital.* **46**:89–122.

Cottier, W. (1953) *Aphids of New Zealand.* Bull. N.Z. Dep. Scient. Ind. Res. 106.

Cuellar, O. (1977) Animal parthenogenesis. *Science* **197**:837–843.

Crawley, M. J. (1983) *Herbivory: The Dynamics of Animal-Plant Interactions.* Blackwell, Oxford.

Davidson, J. (1927) The biological and ecological aspects of migration in aphids. *Sci. Progr. Twent. Cent.* **21**:641–58, **22**:57–69.

Dawkins, R. (1982) *The Extended Phenotype.* Oxford University Press.

DeLoach, C. J. (1974) Rate of increase of populations of cabbage, green peach, and turnip aphids at constant temperatures. *Ann. Ent. Soc. Am.* **67**:332–340.

Dethier, V. G. (1947) *Chemical Insect Attractants and Repellants.* Lewis, London.

Dewar, A. McL. (1977) *Morph determination and host alternation in the apple-grass aphid,* Rhopalosiphum insertum *W alk.* Ph.D. thesis, University of Glassgow.

Dickson, R. C. (1962) Development of the spotted alfalfa aphid population in North America. *Int. Kong. fur Ent.* (*Wien 1960*) **2**:26–28.

Dighton, J. (1978*a*) Effects of synthetic lime aphid honeydew on populations of soil organisms. *Soil Biol. Biochem.* **10**:369–76.

Dighton, J. (1978*b*) *In vitro* experiments simulating the possible fates of aphid honeydew sugars in soil. *Soil. Biol. Biochem.* **10**:53–57.

Dixon, A. F. G. (1958) Escape responses shown by certain aphids to the presence of the coccinellid, *Adalia decempunctata* (L.) *Trans. R. Ent. Soc. Lond.* **10**:319–334.

Dixon, A. F. G. (1969) Population dynamics of the sycamore aphid *Drepanosiphum platanoides* (Schr.) (Hemiptera:Aphididae): migratory and trivial flight activity. *J. anim. Ecol.* **38**:585–606.

Dixon, A. F. G. (1970*a*) 'Quality and availability of food for a sycamore aphid population', in *Animal Populations in Relation to their Food Resources*, ed. A. Watson, Blackwell, Oxford, pp. 271–287.

Dixon, A. F. G. (1970*b*) Stabilization of aphid populations by an aphid induced plant factor. Nature **227**:1368–69.

Dixon, A. F. G. (1971*a*) The 'interval timer' and photoperiod in the determination of parthenogenetic and sexual morphs in the aphid, *Drepanosiphum platanoides. J. Insect Physiol.* **17**:251–60.

Dixon, A. F. G. (1971*b*) Migration in aphids. *Sci. Prog.* **59**:41–53.

Dixon, A. F. G. (1971*c*) The life-cycle and host preferences of the bird cherry-oat aphid, *Rhopalosiphum padi* L., and their bearing on theories of host alternation in aphids. *Ann. appl. Biol.* **68**:135–147.

Dixon, A. F. G. (1971*d*) The role of intra-specific mechanisms and predation in regulating the numbers of the lime aphid *Eucallipterus tiliae* L. *Oecologia (Berl.)* **8**:179–193.

Dixon, A. F. G. (1971*e*) The role of aphids in wood formation. I The effect of the sycamore aphid, *Drepanosiphum platanoides* (Schr.) (Aphididae) on the growth of sycamore, *Acer pseudoplatanus* (L.). *J. appl. Ecol.* **8**:165–79.

Dixon, A. F. G. (1971*f*) The role of aphids in wood formation. II The effect of the lime aphid *Eucallipterus tiliae* L. (Aphididae), on the growth of lime *Tilia* × *vulgaris* Hayne. *J. appl. Ecol.* **8**:393–9.

Dixon, A. F. G. (1972*a*) Crowding and nutrition in the induction of macropterous alatae in *Drepanosiphum dixoni. J. Insect Physiol.* **18**:459–464.

Dixon, A. F. G. (1972*b*) Fecundity of brachypterous and macropterous alatae in *Drepanosiphum dixoni* (Callaphididae, Aphididae). *Entomol. exp. appl.* **15**:335–340.

Dixon, A. F. G. (1972*c*) The 'interval timer' photoperiod and temperature in the seasonal development of parthenogenetic and sexual morphs in the lime aphid, *Eucallipterus tiliae* L. *Oecologia* (*Berl.*) **9**:301–10.

Dixon, A. F. G. (1973) Metabolic acclimatization to seasonal changes in temperature in the sycamore aphid, *Drepanosiphum platanoides* (Schr.) and lime aphid, *Eucallipterus tiliae* L. *Oecologia (Berl.)* **13**: 205–210.

Dixon, A. F. G. (1974*a*) Wing loading and flight activity in the sycamore aphid, *Drepanosiphum plantanoides. Entomol. exp. appl.* **17**:157–162.

Dixon, A. F. G. (1974*b*) Changes in the length of the appendages and the number of rhinaria in young clones of the sycamore aphid, *Drepanosiphum platanoides. Entomol. exp. appl.* **17**:1–8.

Dixon, A. F. G. (1975*a*) 'Aphids and translocation', in *Transport in Plants I. Phloem Transport*, eds M. H. Zimmerman and J. A. Milburn, Springer, Berlin, pp. 154–170.

Dixon, A. F. G. (1975*b*) Seasonal changes in fat content, form, state of gonads and length of adult life in the sycamore aphid, *Drepanosiphum platanoides* (Schr.) *Trans. R. ent. Soc. Lond.* **127**:87–99.

Dixon, A. F. G. (1977) Aphid ecology: life cycles, polymorphism and population regulation. *Ann. Rev. Ecol. Syst.* **8**: 329–353.

Dixon, A. F. G. (1979) 'Sycamore aphid numbers: the role of weather, host and aphid', in *Population Dynamics*, eds. R. M. Anderson, B. D. Turner, and L. R. Taylor, Blackwell, Oxford, pp. 105–121.

Dixon, A. F. G. (1985*a*) 'Parthenogenetic reproduction and the intrinsic rate of increase of aphids', in *Aphids, their Biology, Natural Enemies and Control*, eds. P. Harrewijn and A. Minks, Elsevier, Amsterdam, in press.

Dixon, A. F. G. (1985*b*) 'Adaptive significance of cyclical parthenogenesis in aphids', in *Aphids, their Biology, Natural Enemies and Control*, eds. P. Harrewijn and A. Minks, Elsevier, Amsterdam, in press.

Dixon, A. F. G., Burns, M. D. and Wangboonkong, S. (1968) Migration in aphids: response to current adversity. *Nature* **220**:1337–1338.

Dixon, A. F. G., Chambers, R. J. and Dharma, T. R. (1982) Factors affecting size in aphids with particular reference to the black bean aphid, *Aphis fabae. Entomol. exp. appl.* **32**: 123–128.

Dixon, A. F. G. and Dharma, T. R. (1980*a*) 'Spreading of the risk' in developmental mortality: size, fecundity and reproductive rate in the black bean aphid. *Entomol. exp. appl.* **28**:301–312.

Dixon, A. F. G. and Dharma, T. D. (1980*b*) Number of ovarioles and fecundity in the black bean aphid, *Aphis fabae. Entomol. exp. appl.* **28**:1–14.

Dixon, A. F. G. and Glen, D. M. (1971) Morph determination in the bird cherry-oat aphid, *Rhopalosiphum padi* L. *Ann. appl. Biol.* **68**: 11–21.

Dixon, A. F. G., Kindlemann, P., Lipš, J. and Holman, J. (1985) Why are there so few species of aphids, especially in the tropics? (in press).

Dixon, A. F. G. and Logan, M. (1973) Leaf size and availability of space to the sycamore aphid *Drepanosiphum platanoides. Oikos* **24**: 58–63.

Dixon, A. F. G. and McKay, S. (1970) Aggregation in the sycamore aphid *Drepanosiphum plantanoides* (Schr.) (Hemiptera: Aphididae) and its relevance to the regulation of population growth. *J. anim. Ecol.* **39**:439–454.

Dixon, A. F. G. and Mercer, D. R. (1983) Flight behaviour in the sycamore aphid: Factors affecting take-off. *Entomol. exp. appl.* **33**:43–49.

Dixon, A. F. G. and Stewart, W. A. (1975) Function of the siphunculi in aphids with particular reference to the sycamore aphid, *Drepanosiphum platanoides*. *J. Zool.* (*Lond.*) **175**:279–289.

Dixon, A. F. G. and Wellings, P. W. (1982) Seasonality and reproduction in aphids. *Int. J. inv. Reprod.* **5**:83–89.

Dixon, A. F. G. and Wratten, S. D. (1971) Laboratory studies on aggregation, size and fecundity in the black bean aphid, *Aphis fabae* Scop. *Bull. ent. Res.* **61**:97–111.

Downing, N. (1978) Measurements of the osmotic concentrations of stylet sap, haemolymph and honeydew from an aphid under osmotic stress. *J. exp. Biol.* **77**:247–250.

Doyle, R. W. and Hunte, W. (1981) Demography of an estuarine Amphipod (*Gammarus lawrencianus*) experimentally selected for high "r": A model of the genetic effects of environmental change. *Can. J. Fish. Aquat. Sci.* **38**:1120–1127.

Dritschilo, W., Krummel, J., Nafus, D. and Pimentel, D. (1979) Herbivorous insects colonising cyanogenic and acyanogenic *Trifolium repens*. *Heredity* **42**:49–56.

Duckett, D. P. (1974) *Further studies of ant-aphid interactions*. Ph.D. thesis, University of London.

Dunn, J. A. (1959) The survival in soil of apterae of lettuce root aphid. *Ann. appl. Biol.* **47**:766–771.

Dunn, J. A. (1960) The formation of galls by some species of *Pemphigus* (Homoptera: Aphididae). *Marcellia* **30**:155–167.

Eastop, V. F. (1958) *A Study of the Aphididae (Homoptera) of East Africa*. HMSO, London.

Eastop, V. F. (1961) *A Study of the Aphididae (Homoptera) of West Africa*. HMSO, London.

Eastop, V. F. (1966) A taxonomic study of Australian Aphidoidea (Homoptera). *Aust. J. Zool.* **14**:399–592.

Eastop, V. F. (1973) 'Deductions from the present day host plants of aphids and related insects', in *Insect/Plant Relationships, Symp. R. ent. Soc. Lond.* **6**:157–178.

Eastop, V. F. (1977) 'Worldwide importance of aphids as virus vectors', in *Aphids as Virus Vectors*, eds. K. F. Harris and K. Maramorosch, Academic Press, London, pp. 3–62.

Ehrhardt, P. (1968) Einfluss von Ernährungsfaktoren auf die Entwicklung von säfte saugenden Insekten unter besonderer Berücksichtigung von Symbionten. *Z. Parasitenk.* **31**:38–66.

El-Shazly, N. Z. (1972) Der Einfluss äusserer Faktoren auf die hämocytäre Abwehrreaktion von *Neomyzus circumflexus* (Buck.) (Homoptera: Aphididae). *Z. ang. Ent.* **70**:414–436.

Elton, C. S. (1925) The dispersal of insects to Spitsbergen. *Trans. Ent. Soc. Lond.* (1925):289–299.

El-Ziady, S. (1960) Further effects of *Lasius niger* L. on *Aphis fabae* Scopoli. *Proc. R. Ent. Soc. Lond* (*A*) **35**:30–38.

Eschrich, W. (1970) Biochemistry and fine structure of phloem in relation to transport. *Ann. Rev Plant Physiol.* **21**:193–214.

Ewing, H. E. (1916) Eighty-seven generations in a parthenogenetic pure line of *Aphis avenae* Fab. *Biol. Bull.* **31**:53–112.

Fenchel, T. (1974) Intrinsic rate of natural increase: the relationship with body size. *Oecologia (Berl.)* **14**:317–326.

Fisher, M. (1982) *Morph determination in* Elatobium abietinum *(Walk.) the green spruce aphid*. Ph.D. thesis, University of East Anglia.

Fisher, R. A. (1930) *The Genetical Theory of Natural Selection*. Oxford University Press.

Forbes, A. R. (1966) Electron microscope evidence for nerves in the mandibular stylets of the green peach aphid. *Nature* **212**:726.

Forrest, J. M. S. (1970) The effects of maternal and larval experience on morph determination in *Dysaphis devecta*. *J. insect Physiol.* **16**:2281–92.

Forrest, J. M. S. (1971) The growth of *Aphis fabae* as an indicator of the nutritional advantages of galling to the apple aphid *Dysaphis devecta*. *Entomologia exp. et appl.* **14**:477–483.

Forrest, J. M. S. and Dixon, A. F. G. (1975) The induction of leaf-roll galls by the apple aphids *Dysaphis devecta* and *D. plantaginea Ann. appl. Biol.* **81**:281–288.

Foster, W. A. (1978) Dispersal behaviour of an intertidal aphid. *J. anim. Ecol.* **47**:653–659.

Foster, W. A. and Treherne, J. E. (1978) Dispersal mechanisms in an intertidal aphid. *J. anim. Ecol.* **47**:205–217.

Frank, A. B. (1896) *Die Krankheiten der Pflanzen Bd. III: Die tierparasitären Krankheiten der Pflanzen*. 2nd edn. Ed. Trewendt, Breslau.

Frazer, B. D. and Gill, B. (1981) Age, fecundity, weight, and the intrinsic rate of increase of the lupine aphid *Macrosiphum albifrons* (Homoptera: Aphididae). *Can. Ent.* **113**:739–745.

Fretwell, S. D. and Lucas, H. L. (1969) On territorial behaviour and other factors influencing habitat distribution in birds. I. Theoretical development. *Acta Biotheoretica* **19**:16–36.

de Geer, C. (1773) *Mémoires pour servir a l'Histoire des Insectes: Trois Mem. Suite des Pucerons.* P. Hasselberg, Stockholm.

Geiler, H. (1956) Eichhörnchen, *Sciurus vulgaris fuscoater* Altum, 1876, als Vertilger von Blattläusen (Aphidae). *Saugentierkundl. Mitt.* **4**:13–15.

Ghosh, A. K. (1980) *The Fauna of India and the Adjacent Countries. Homoptera, Aphidoidea Part I. General Introduction and Chaitophorinae.* Zoological Survey of India, Madras.

Gibson, R. W. and Pickett, J. A. (1983) Wild potato repels aphids by release of aphid alarm pheromone. *Nature* **302**:608–609.

Gilbert, N. (1980) Comparative dynamics of a single-host aphid II. Theoretical consequences. *J. anim. Ecol.* **49**:371–380.

Grubb, P. J., Kelly, D. and Mitchley, J. (1982) 'The control of relative abundance in communities of herbaceous plants', in *The Plant Community as a Working Mechanism*, ed. E. I. Newman, Blackwell, Oxford, pp. 79–97.

Haine, E. (1955) Aphid take-off in controlled wind speeds. *Nature* **175**:474.

Hales, D. F. (1976) Biology of *Neophyllaphis brimblecombei* Carver (Homoptera: Aphididae) in the Sydney Region. *Aust. Zool.* **19**:77–83.

Hales, D. F. and Mittler, T. E. (1983) Precocene causes male determination in the aphid *Myzus persicae. J. Insect Physiol.* **29**:819–823.

Hamilton, W. D. (1967) Extraordinary sex ratios. *Science* **156**:477–488.

Hanski, I. (1980) Spatial patterns and movements in coprophagous beetles. *Oikos* **34**:293–310.

Hanski, I. (1982) On patterns of temporal and spatial variation in animal populations. *Ann. Zool. Fennici* **19**:21–37.

Harris, K. F. (1977) 'An ingestion-esgestion hypothesis of non circulative virus transmission', in *Aphidis as Virus Vectors*, eds. K. F. Harris and K. Maramorosch, Academic Press, New York, pp. 165–220.

Heathcote, G. D. (1974) Aphids caught on sticky traps in eastern England in relation to the spread of yellowing viruses of sugar beet. *Bull. ent. Res.* **64**:669–76.

Heie, O. E. (1967) Studies on fossil aphids (Homoptera: Aphidoidea). *Spolia Zool. Mus. Hauniensis* **26**:7–274.

Heie, O. E. (1980) *The Aphidoidea (Hemiptera) of Fennoscania and Denmark I. Fauna Ent. Scand.* (9), 236 pp.

Heie, O. E. (1982) *The Aphidoidea (Hemiptera) of Fennoscandia and Denmark II. Fauna Ent. Scand.* (11), 176 pp.

Hennig, E. (1966) Zur Histologie und Funktion von Einstichen der Schwarzen Bohnenlaus (*Aphis fabae* Scop.) in *Vicia faba*—Pflanzen. *J. Insect Physiol.* **12**:65–76.

Herger, P. (1975) Einfluss von Farbe und Nahrungszusammensetzung auf das Saugverhalten der kunstlich ernahrten Ampferblattlaus *Aphis rumicis* (Homoptera: Aphididae). *Ent. Ger.* **2**:149–166.

Higuchi, H. (1972) A taxonomic study of the sub-family Callipterinae in Japan. *Insecta Matsumurana* **35**(2):19–126.

Hille Ris Lambers, D. (1938) Contributions to a monograph of the Aphididae of Europe. *I. Temminckia* **3**:1–44.

Hille Ris Lambers, D. (1939) Contributions to a monograph of the Aphididae of Europe. II. *Temminckia* **4**:1–134.

Hille Ris Lambers, D. (1945) De Bloedulekkenluis van appel, *Sappahis devecta* (Walker). *Tidschr. Plz.* **51**:57–72.

Hille Ris Lambers, D. (1947) Contributions to a monograph of the Aphididae of Europe. III. *Temminckia* **7**:179–319.

Hille Ris Lambers, D. (1949) Contributions to a monograph of the Aphididae of Europe. IV. *Temminckia* **8**:182–323.

Hille Ris Lambers, D. (1953) Contributions to a monograph of the Aphididae of Europe. V. *Temminckia* **9**:1–176.

Holman, J. (1971) 'Factors influencing the host range of the aphids in the tropics', *Proc. 13th Int. Cong. Ent. (Moscow)* **2**:339–340.

Houk, E. J. and Griffiths, G. W. (1980) Intracellular symbiotes of the Homoptera. *Ann. Rev. Entomol.* **25**:161–87.

Houk, E. J., Griffiths, G. W. and McLean, D. L. (1976) Lipid metabolism in the symbiotes of the pea aphid *Acyrthosiphon pisum*. *Comp. Biochem. Physiol.* **54B**:427–431.

Huxley, T. H. (1858) On the agamic reproduction and morphology of *Aphis*—Part 1. *Trans. Linn. Soc.* **22**:193–219.

Ibbotson, A. and Kennedy, J. S. (1950) The distribution of aphid infestation in relation to leaf age II. The progress of *Aphis fabae* Scop. infestations on sugar beet in pots. *Ann. appl. Biol.* **37**:680–696.

Itô, Y. (1980) *Comparative Ecology*. Cambridge University Press.

James, B. D. and Luff, M. L. (1982) Cold-hardiness and development of eggs of *Rhopalosiphum insertum*. *Ecol. Ent.* **7**:277–282.

Janzen, D. H. (1977) What are dandelions and aphids? *Amer. Nat.* **111**:586–589.

Johnson, B. (1958) Factors affecting the locomotor and settling responses of alate aphids. *Anim. Behav.* **6**:9–26.

Johnson, C. G. (1969) *Migration and Dispersal of Insects by Flight*. Methuen, London.

Johnson, C. G. Taylor, L. R. and Haine, E. (1957) The analysis and reconstruction of diurnal flight curves in alienicolae of *Aphis fabae* Scop. *Ann. appl. Biol.* **45**:682–701.

Jones, M. G. (1942*a*) A description of *Aphis (Doralis) rumicis*, L., and comparison with *Aphis (Doralis) fabae*, Scop. *Bull ent. Res.* **33**:5–20.

Jones, M. G. (1942*b*) The summer hosts of *Aphis fabae* Scop. *Bull. Entomol. Res.* **33**: 161–69.

Judge, F. D. (1967) Overwintering in *Pemphigus bursarius* (L.) *Nature* **216**:1041–1042.

Kennedy, G. G. and Kishaba, A. N. (1977) Response of alate melon aphids to resistant and susceptible muskmelon lines. *J. Econ. Ent.* **70**:407–410.

Kennedy, J. S. (1975) 'Insect dispersal', in *Insects, Science and Society*, Conf. Proc., ed. D. Pimentel, Academic Press, New York etc., pp. 103–119.

Kennedy, J. S. (1976) Host-plant finding by flying aphids. *Symp. Biol. Hung.* **16**:121–123.

Kennedy, J. S. and Booth, C. D. (1951) Host alternation in *Aphis fabae* Scop. 1. Feeding preferences and fecundity in relation to age and kind of leaves. *Ann. appl. Biol.* **38**:25–64.

Kennedy, J. S. and Booth, C. D. (1963*a*) Free flight of aphids in the laboratory. *J. exp. Biol.* **40**:67–85.

Kennedy, J. S. and Booth, C. D. (1963*b*) Co-ordination of successive activities in an aphid. The effect of flight on the settling responses. *J. exp Biol.* **40**:351–369.

Kennedy, J. S., Booth, C. O. and Kershaw, W. J. S. (1959) Host finding by aphids in the field. I. Gynoparae of *Myzus persicae* (Sulzer). *Ann. appl. Biol.* **47**:410–423.

Kennedy, J. S., Booth, C. O. and Kershaw, W. J. S. (1961) Host finding by aphids in the field. III. Visual attraction. *Ann. app. Biol.* **49**:1–21.

Kennedy, J. S., Ibbotson, A. and Booth, C. O. (1950) The distribution of aphid infestation in relation to leaf age. I. *Myzus persicae* (Sulz.) and *Aphis fabae* Scop. on spindle trees and sugar-beet plants. *Ann. appl. Biol.* **37**:651–679.

Kennedy, J. S. and Stroyan, H. L. G. (1959) Biology of aphids. *Ann. rev. Entomol.* **4**: 139–160.

Kiss, A. (1981) Melezitose, aphids and ants. *Oikos* **37**:382.

Klingauf, F. (1971) Die Wirkung des Glucosids Phlorizin auf das Wirtswahlverhalten von *Rhopalosiphum insertum* (Walk.) und *Aphis pomi* De Geer (Homoptera:Aphididae). *Zeit. Angew. Ent.* **68**:41–55.

Klingauf, F. (1972) Die Bedeutung von peripher vorliegenden Pflanzen substanzen für die Wirtswahl von phloemsaugenden Blattläusen (Aphididae). *Z. Pflanzenkrankh. Pflanzenschutz.* **79**:471–477.

Kloft, W. J. (1977) 'Radioisotopes in aphid research', in *Aphids as Virus Vectors*, eds. K. F. Harris and K. Maramorosch, Academic Press, New York etc., pp. 291–310.

Knight, R. L. and Alston, F. H. (1974) 'Pest resistance in fruit breeding', in *Biology in Pest and Disease Control*, eds. D. Price Jones and M. E. Solomon, Blackwell, Oxford, pp. 73–86.

Kunkel, H. and Mittler, T. E. (1971) Einfluss der Ernährung bei Junglarven von *Myzus persicae* (Sulz.) (Aphididae) auf ihre Entwicklung zu Geflügelten oder Ungeflügelten. *Oecologia (Berl.)* **8**:110–134.

Lamb, R. J. and Mackay, P. A. (1979) Variability in migratory tendency within and among natural populations of the pea aphid, *Acyrthosiphon pisum*. *Oecologia (Berl.)* **39**:289–299.

Lamb, R. J. and Mackay, P. A. (1983) 'Micro-evolution of the migratory tendency, photoperiodic response and developmental threshold of the pea aphid, *Acyrthosiphon pisum*', in *Diapause and Life Cycle Strategies in Insects*, eds. V. K. Brown and I. Hodek, Dr. W. Junk, The Hague, pp. 209–217.

Lawson, C. A. (1941) The effect of temperature on longevity, reproduction and growth in aphids. *Genetics* **26**:159.

Leather, S. R. (1980*a*) Egg survival in the bird cherry-oat aphid, *Rhopalosiphum padi* (L.) *Entomol. exp. appl.* **27**:96–97.

Leather, S. R. (1980*b*) *Aspects of the ecology of the bird cherry-oat aphid* Rhopalosiphum padi (*L.*) Ph.D. thesis, University of East Anglia.

Leather, S. R. (1981) Factors affecting egg survival in the bird cherry-oat aphid, *Rhopalosiphum padi. Entomol. exp. appl.* **30**:197–199.

Leather, S. R. and Dixon, A. F. G. (1984) Aphid growth and reproductive rates. *Entomol. exp. appl.* **35**:137–140.

Leather, S. R., Ward, S. A. and Dixon, A. F. G. (1983) The effect of nutrient stress on some life history parameters of the black bean aphid, *Aphis fabae* Scop. *Oecologia (Berl.)* **57**:156–7.

Leather, S. R. and Wellings, P. W. (1981) Ovariole number and fecundity in aphids. *Entomol. exp. appl.* **30**:128–133.

Lees, A. D. (1959) The role of photoperiod and temperature in the determination of parthenogenetic and sexual forms in the aphid *Megoura viciae* Buckton I. The influence of these factors on apterous virginoparae and their progeny. *J. Insect Physiol.* **3**:92–117.

Lees, A. D. (1960) The role of photoperiod and temperature in the determination of parthenogenetic and sexual forms in the aphid *Megoura viciae* Buckton. II The operation of the 'internal timer' in young clones. *J. Insect Physiol.* **4**:154–175.

Lees, A. D. (1964) The location of the photoperiodic receptors in the aphid *Megoura viciae* Buckton. *J. exp. Biol.* **41**:119–133.

Lees, A. D. (1966) The control of polymorphism in aphids. *Adv. Insect Physiol.* **3**:207–277.

Lees, A. D. (1967) The production of the apterous and alate forms in the aphid *Megoura viciae* Buckton, with special reference to the role of crowding. *J. Insect Physiol.* **13**:289–318.

Lees, A. D. (1973) Photoperiodic time measurement in the aphid *Megoura viciae. J. Insect Physiol.* **19**:2279–2316.

Lees, A. D. (1979) 'The maternal environment and the control of morphogenesis in insects', in *Maternal Effects in Development*, eds. D. R. Newth and M. Balls, Brit. Soc. Develop. Biol. Symp. (4), pp. 221–239.

Lees, A. D. and Hardie, J. (1981) 'The photoperiodic control of polymorphism in aphids: neuroendocrine and endocrine components', in *Biological Clocks in Seasonal Reproductive Cycles*, eds. B. K. Follett and D. E. Follett, Scientechnica, Bristol, pp. 125–135.

Llewellyn, M. J. (1970) *The ecological energetics of the lime aphid* Eucallipterus tiliae *(L.) and its effect on tree growth*. Ph.D. thesis, University of Glasgow.

Llewellyn, M. (1972) The effect of the lime aphid, *Eucallipterus tiliae* L. (Aphididae) on the growth of the lime *Tilia* × *vulgaris* Hayne. I. Energy requirements of the aphid populations. *J. appl. Ecol.* **9**:261–82.

Llewellyn, M. (1975) The effects of the lime aphid (*Eucallipterus tiliae* L) (Aphididae) on the growth of the lime *Tilia* × *vulgaris* Hayne) II. The primary production of saplings and mature trees, the energy drain imposed by the aphid populations and revised standard deviations of aphid population energy budgets. *J. appl. Ecol.* **12**:15–23.

Llewellyn, M. (1982) 'The energy economy of fluid-feeding herbivorous insects', in *Proc. 5th Int. Symp. Insect-Plant Relationships, Wageningen*, Pudoc, Wageningen, pp. 243–251.

Llewellyn, M., Rashid, R. and Leckstein, P. (1974) The ecological energetics of the willow aphid *Tuberolachnus salignus* (Gmelin): honeydew production. *J. Anim. Ecol.* **43**:19–29.

Long, B. J., Dunn, G. M., Bowman, J. S. and Routley, D. G. (1977) Relationship of hydroxamic acid content in corn and resistance to the corn leaf aphid. *Crop Sci.* **17**:55–58.

Macfadyen, A. (1964) 'Energy flow in ecosystems and its exploitation by grazing', in *Grazing in Terrestrial and Marine Environments*, ed. D. J. Crisp, Blackwell, Oxford, pp. 3–20.

MacGibbon, D. B. (1975) An improved reagent for detecting glucosinolase activity in acrylamide gel electropherograms. *N.Z. J. Sci.* **18**:217–219.

MacGibbon, D. B. and Benzenberg, E. J. (1978) Location of glucosinolase in *Brevicoryne brassicae* and *Lipaphis erysimi* (Aphididae). N.Z. J. Sci. **21**:389–92.

Mackauer, M. (1973) 'Host selection and host suitability in *Aphidius smithi*', in *Perspectives in Aphid Biology*, ed. A. D. Lowe, Bull. Ent. Soc. N.Z. **2**:20–29.

Marcovitch, S. (1924) The migration of the Aphididae and the appearance of the sexual forms as affected by the relative length of daily light exposure. *J. Agr. Res.* **27**:513–522.

Markkula, M. (1963) Studies on the pea aphid, *Acyrthosiphon pisum* Harris (Hom., Aphididae) with special reference to the difference in biology of the green and red forms. *Ann. Agr. Fenn.* **2**: Suppl. **1**: 1–30.

Marthavan, S. and Pandian, T. J. (1975) Effect of temperature on food utilization in the monarch butterfly *Danaus chrysippus*. *Oikos* **26**:60–64.

Massee, A. M. (1946) *The Pests of Fruits and Hops*. Crosby Lockwood, London.

Matsuka, M. and Mittler, T. E. (1978) Enhancement of alata production by an aphid, *Myzus persicae*, in response to increases in daylength. *Bull. Fac. Agric. Tamagawa Univ. Tokyo* **18**: 1–7.

Maynard Smith, J. (1978) *The Evolution of Sex*. Cambridge University Press.

Maynard Smith, J. (1984) 'The ecology of sex', in *Behavioural Ecology: An Evolutionary Approach*, eds. J. R. Krebs and N. B. Davis, 2nd edn., Blackwell, Oxford, pp. 201–221.

McLean, D. L. and Kinsey, M. G. (1968) Probing behaviour of the pea aphid, *Acyrthosiphon pisum*. II. Comparisons of salivation and ingestion in host and non-host leaves. *Ann. ent. Soc. Am.* **61**: 730–39.

Mercer, D. R. (1979) *Flight behaviour of the sycamore aphid* Drepanosiphum platanoidis *Schr.* Ph.D. thesis, University of East Anglia.

Michel, E. (1942) Beiträge zur Kenntnis von *Lachnus (Pterochlorus) roboris* L., einer wichtigen Honigtauerzeugerin an der Eiche. Z. *Angew. Ent.* **29**:243–281.

Miles, P. W. (1968*a*) Studies on the salivary physiology of plant-bugs: Experimental induction of galls. *J. Insect Physiol.* **14**:97–106.

Miles, P. W. (1968*b*) Insect secretions in plants. *Rev. Phytopathol.* **6**: 137–164.

Miles, P. W. (1978) Redox reactions of hemipterous saliva in plant tissues. *Ent. exp. Appl.* **24**: 334–39.

Mittler, T. E. (1957) Studies on the feeding and nutrition of *Tuberolachnus salignus* (Gmelin) (Homoptera, Aphididae). I. The uptake of phloem sap. *J. exp. Biol.* **34**: 334–341.

Mittler, T. E. (1958*a*) Studies on the nutrition of *Tuberolachnus salignus* (Gmelin) (Homoptera, Aphididae). II. The nitrogen and sugar composition of ingested phloem sap and excreted honeydew. *J. exp. Biol.* **35**: 74–84.

Mittler, T. E. (1958*b*) Studies on the nutrition of *Tuberolachnus salignus* (Gmelin) (Homoptera, Aphididae). III. The nitrogen economy. *J. exp. Biol.* **35**:626–638.

Mittler, T. E. (1971) Some effects on the aphid *Myzus persicae* of ingesting antibiotics incorporated into artificial diets. *J. Insect Physiol.* **17**: 1333–1347.

Mittler, T. E. and Kleinjan, J. E. (1970) Effect of artificial diet composition on wing-production by the aphid *Myzus persicae*. *J. Insect Physiol.* **16**: 833–850.

Miyazaki, M. (1971) A revision of the tribe Macrosiphini of Japan (Homoptera: Aphididae, Aphidinae). *Insecta Matsumurana* **34**: 1–247.

Mordvilko, A. K. (1908) Beitrage zur Biologie der Pflanzenläuse, *Aphididae* Passerini. *Biol. Zbl.* **28**: 631–639.

Mordvilko, A. K. (1928) The evolution of cycles and the origin of heteroecy (migrations) in plant-lice. *Ann. Mag. Nat. Hist. (Ser. 10)* **2**:570–582.

Mordvilko, A. K. (1935) Aphids: their generation cycles and evolution. *Priroda* **11**: 35–44.

Müller, F. P. (1962) Biotypen and Unterarten der "Erbsenlaus" *Acyrthosiphon pisum* Harris. *Z. Pflkrankh. Pflschutz.* **69**: 129–136.

Müller, H. J. (1964) The relation of recombination to mutational advance. *Mutat. Res.* **1**:2–9.

Müller, H. J. (1966*a*) Über die Ursachen des unterschiedlichen Resistenz von *Vicia faba* L. gegen über der Bohnenblattlaus. *Aphis (Doralis) fabae* Scop. IX. Der Einfluss okologischer Faktoren auf das Wachstum von *Aphis fabae* Scop. *Entomol. exp. appl.* **9**: 42–66.

Müller, H. J. (1966*b*) Über Mehrjährige Coccinelliden-Fänge auf Ackerbohnen mit hohem *Aphis fabae*—Besatz. *Z. Morphol. Oekol. Tiere* **58**: 144–161.

Murdie, G. (1969*a*) Some causes of size variation in the pea aphid, *Acyrthosiphon pisum* Harris. *Trans. Roy. entomol. Soc. Lond.* **121**:423–442.

Murdie, G. (1969*b*) The biological consequences of decreased size caused by crowding or rearing temperatures in apterae of the pea aphid, *Acyrthosiphon pisum* Harris. *Trans. Roy. entomol. Soc. Lond.* **121**: 443–455.

Nault, L. R. and Montgomery, M. E. (1977) 'Aphid pheromones', in *Aphids as Virus Vectors*, eds. K. F. Harris and K. Maramorosch, Academic Press, New York etc., pp. 527–545.

Nault, L. R. and Styer, W. E. (1972) Effects of sinigrin on host selection by aphids. *Entomol. exp. appl.* **15**: 423–37.

Nielson, M. W. and Don, H. (1974) Probing behaviour of biotypes of the spotted alfalfa aphid on resistant and susceptible alfalfa clones. *Entomol. exp. appl.* **17**: 477–486.

Noda, I. (1960) The emergence of winged viviparous female in aphid. VI. Difference in the rate of development between the winged and the unwinged forms. *Jap. J. Ecol.* **10**:97–102.

Orlando, E. (1974) Sex determination in *Megoura viciae* Buckton (Homoptera, Aphididae) *Monitore zool. ital. (N.S.)* **8**:61–70.

Owen, D. F. (1978) Why do aphids synthesize melezitose? *Oikos* **31**: 264–267.

Owen, D. F. (1980*a*) How plants may benefit from the animals that eat them? *Oikos* **35**:230–35.

Owen, D. F. (1980*b*) Response to Petelle's comments. *Oikos* **35**:128.

Owen, D. F. and Wiegert, R. G. (1976) Do consumers maximize plant fitness? *Oikos* **27**:488–492.

Owen, D. F. and Wiegert, R. G. (1981) Mutualism between grasses and grazers: an evolutionary hypothesis. *Oikos* **36**: 376–378.

Paik, W. H. (1965) *Aphids of Korea* Seoul National University.

Paik, W. H. (1972) *Illustrated Encyclopedia of Fauna and Flora of Korea* 13: Insecta 5. (In Korean).

Palmer, M. A. (1952) *Aphids of the Rocky Mountain Region*, Vol. 5. The Thomas Soy Foundation.

Parrish, W. B. (1967) The origin, morphology, and innervation of aphid stylets (Homoptera) *Ann. entomol. Soc. Am.* **60**:273–276.

Parry, W. H. (1979*a*) Factors affecting low temperature survival of *Cinara pilicornis* eggs on Sitka spruce. *Int. J. Biometeorology* **23**:185–193.

Parry, W. H. (1979*b*) Acclimatization in the green spruce aphid, *Elatobium abietinum. Ann. appl. Biol.* **92**:299–306.

Petelle, M. (1980) Aphids and melezitose: a test of Owen's 1978 hypothesis. *Oikos* **35**:127–8.

Pettersson, J. (1970) Studies on *Rhopalosiphum padi* (L.) I. Laboratory studies on olfactometric responses to the winter host, *Prunus padus. Lantbrukshogskolans Annaler* **36**:381–399.

Pickett, J. A. and Griffiths, D. C. (1980) Composition of aphid alarm pheromones. *J. chem. Ecol.* **6**:349–360.

Pollard, D. G. (1973) Plant penetration by feeding aphids (Hemiptera, Aphidoidea): a review. *Bull. ent. Res.* **62**:631–714.

Pollard, D. G. (1977) 'Aphid penetration of plant tissues', in *Aphids as Virus Vectors*, eds. K. F. Harris and K. Maramorosch, Academic Press, New York etc., pp. 105–118.

Powell, W. (1974) Supercooling and the low-temperature survival of the green spruce aphid *Elatobium abietinum. Ann. appl. Biol.* **78**:27–37.

Powell, W. and Parry, W. H. (1976) Effects of temperature on overwintering populations of the green spruce aphid *Elatobium abietinum. Ann. appl. Biol.* **82**: 209–219.

Rainey, R. C. (1963) Meterology and the migrations of desert locusts—applications of synoptic meteorology in locust control. *Anti-locust Memoir* **7**.

Richards, W. R. (1960) A synopsis of the genus *Rhopalosiphum* Koch in Canada. *Can. Ent.* **92**:1–51.

Richards, W. R. (1963) The myzaphidines of Canada (Homoptera:Aphididae) *Can. Ent.* **95**:680–704.

Richards, W. R. (1965) *The Callaphidini of Canada (Homoptera: Aphididae)* Mem. Ent. Soc. Canada No. 44.

Richards, W. R. (1972) *The Chaitophorinae of Canada (Homoptera: Aphididae)*. Mem. Ent. Soc. Canada No. 87.

Roberti, D. (1939) Contributi alla conoscenza degli afidi d'Italia II. *Bull del Lab. Zool. Gen. Agria, Facult. Agraria, Porticii* **31**:137–157.

Rohfritsch, O. (1966) Dehiscence de la Galle d'*Adelges abietis* Kalt. *Marcellia* **33**:149–158.

Rothschild, M. J., von Euw. J., and Reichstein, T. (1970) Cardiac glycosides in the Oleander aphid *Aphis nerii. J. Insect Physiol.* **16**:1141–1145.

Salyk, R. P. and Sullivan, D. J. (1982) Comparative feeding behaviour of two aphid species: bean aphid (*Aphis fabae* Scopoli) and pea aphid (*Acyrthosiphon pisum* (Harris)) (Homoptera:Aphididae) *J. New York Ent. Soc.* **90**:87–93.

Sanders, C. J. and Knight, F. B. (1968) Natural regulation of the aphid *Pterocomma populifoliae* on the big-tooth aspen in northern lower Michigan. *Ecology* **49**: 234–244.

Schaefers, G. A. and Judge, F. D. (1971) Effects of temperature, photo-period and host plant on alary polymorphism in the aphid, *Chaetosiphon fragaefolii. J. Insect Physiol.* **17**:365–379.

Schwarzbach, E. (1962) Die Bildung von Fluglöchern an Gallen von *Pemphigus spirothecae* Pass. *Naturwissenschaften* **49**:91.

Sen Gupta, G. C. and Miles, P. W. (1975) Studies on the susceptibility of varieties of apple to the feeding of two strains of woolly aphis (Homoptera) in relation to the chemical content of the tissues of the host. *Aust. J. agric. Res.* **26**:157–68.

Shambaugh, G. F., Frazier, J. L., Castell, A. E. M. and Coons, L. B. (1978) Antennal sensilla of seventeen aphids (Homoptera: Aphidinae). *Int. J. insect Morph & Embryol.* **7**:389–404.

Shaposhnikov, G. Kh (1964) 'Suborder Aphidinae—plant lice', in *Keys to the Insects of the European USSR*, G. Ya. Bei-Benko *et al.*, (in Russian, English translation, Jerusalem 1967) pp. 616–799.

Shaposhnikov, G. Kh (1977) The trend of evolution. *Zhurnal Obshchei Biologii* **38**:649–655.

Shaposhnikov, G. Kh (1981) *Populations and Species in Aphids and the Need for a Universal Species Concept*. Special Publication, Research Branch, Agriculture Canada.

Shaw, M. J. P. (1970*a*) Effects of population density on alienicolae of *Aphis fabae* Scop. I. The effect of crowding on the production of alatae in the laboratory. *Ann. appl. Biol.* **65**:191–196.

Shaw, M. J. P. (1970*b*) Effect of population density on alienicolae of *Aphis fabae* Scop. II. The effects of crowding on the expression of mgratory urge in the laboratory. *Ann. appl. Biol.* **69**:197–203.

Shaw, M. J. P. (1970) Effect of population density on alienicolae of *Aphis fabae* Scop. III. The effect of isolation on the development of form and behaviour of alatae in a laboratory clone. *Ann. appl. Biol.* **65**:205–212.

Shearer, J. W. (1976) *Polymorphism and population ecology of the European maple aphid, Periphyllus testudinaceus (Fernie)* Unpublished Ph.D. thesis, University of Glasgow.

Shinji, O. (1941) *Monograph of Japanese Aphis* (in Japanese). Tokyo.

Skinner, G. J. and Whittaker, J. E. (1981) An experimental investigation of inter-relationships between the wood-ant (*Formica rufa*) and some tree-canopy herbivores. *J. anim. Ecol.* **50**:313–326.

Sömme, L. (1969) Mannitol and glycerol in overwintering aphid eggs. *Norsk ent. Tidsskr.* **16**:107–111.

Stearns, S. C. (1976) Life-history tactics: a review of the ideas. *Quart. Rev. Biol.* **51**:3–47.

Steel, C. G. H. and Lees, A. D. (1977) The role of neurosecretion in the photoperiodic control of polymorphism in the aphid *Megoura viciae*. *J. exp. Biol.* **67**:117–135.

Stroyan, H. L. G. (1977) *Homoptera Aphidoidea: Chaitophoridae and Callaphidae*. Roy. Ent. Soc. (Lond.) Handbooks for the Identification of British Insects, Vol. 2, Part 4a.

Suomalainen, E. (1950) Parthenogenesis in animals. *Adv. Genet.* **3**:193–253.

Suomalainen, E., Saura, A., Lokki, J. and Teeri, T. (1980) Genetic polymorphism and evolution in parthenogenetic animals. Pt. 9: Absence of variation within parthenogenetic aphid clones. *Theor. appl. Genet.* **57**:129–132.

Sutherland, O. R. W. (1968) Dormancy and lipid storage in the pemphigine aphid *Thecabius affinis*. *Entomol. exp. appl.* **11**:348–354.

Sutherland, O. R. W. (1969*a*) The role of crowding in the production of winged forms by two strains of the pea aphid, *Acyrthosiphon pisum*. *J insect Physiol.* **15**:1385–1410.

Sutherland, O. R. W. (1969*b*) The role of the host plant in the production of winged forms by two strains of the pea aphid, *Acyrthosiphon pisum*. *J. insect Physiol.* **15**:2179–2201.

Sutherland, O. R. W. and Mittler, T. E. (1971) Influence of diet composition and crowding on wing production by the aphid *Myzus persicae*. *J. insect Physiol.* **17**:321–328.

Szelegiewicz, H. (1978) Różnodomnść (heteroecja) u mszyc, jej pochodzenie i ewolucja. *Zesz. Probl. Post Nauk Roln.* **208**:19–31.

Taimr, L., Kudelová, A. and Kříž, J. (1978) Diurnal periodicity in the flight activity of migrant alatae of *Phorodon humuli* Schrank (Hom. Aphididae). *Zeit. ang. Ent.* **36**:373–380.

Takada, H. (1979) Characteristics of forms of *Myzus persicae* (Sulzer) (Homotera: Aphididae) distinguished by colour and esterase differences and their occurrence in populations on different host plants in *Japan*. *Appl. ent. Zool.* **14**:370–375.

Takada, H. (1981) Inheritance of body colors in *Myzus persicae* (Sulzer) (Homoptera: Aphididae). *Appl. ent. Zool.* **16**:242–246.

Tamaki, G. and Allen, W. W. (1969) Competition and other factors influencing the population dynamics of *Aphis gossypii* and *Macrosiphoniella sanborni* on greenhouse chrysanthemums. *Hilgardia* **39**:447–505.

Tarn, T. R. and Adams, J. B. (1982) 'Aphid probing and feeding, electronic monitoring, and plant breeding', in *Pathogens, Vectors and Plant Diseases: Approach to control*, eds. K. F. Harris and K. Maramorosch, Academic Press, New York etc., pp. 221–246.

Tashev, D. and Markova, E. (1982) Morphofunktionellc Grundlagen der Fruchtbarkeit bei den Blattläusen (Aphidina, Viviovipara). I. Anzahl Der Eirohren. *Godishnik Na Sofiiskiya Universitet Kliment Okhridski Biologicheski Fakultet (Zoologiya)* **70**: 19–28

Taylor, L. R. (1961) Aggregation, variance and the mean. *Nature* **189**:732–735.

Taylor, L. R. (1965) Flight behaviour and aphid migration. *Proc. N. Cent. Br. Entomol. Soc. Amer.* **20**:9–19.

Taylor, L. R. (1973) Monitoring change in the distribution and abundance of insects. *Rep. Rothamsted exp. Stn. 1973*:202–239.

Taylor, L. R. (1974) Insect migration, flight periodicity and the boundary layer. *J. anim. Ecol.* **43**:225–238.

Taylor, L. R. (1979) 'The Rothamsted Insect Survey—an approach to the theory and practice of synoptic pest forecasting in agriculture', in *Movement of Highly Mobile Insects*, eds. R. L. Rabb and G. G. Kennedy, pp. 148–185.

Taylor, L. R. and Taylor, R. A. J. (1977) Aggregation, migration and population mechanics. *Nature* **265**:415–421.

Taylor, L. R. and Taylor, R. A. J. (1978) 'The dynamics of spatial behaviour', in *Population Control by Social Behaviour*, eds. F. J. Ebling and D. M. Stoddart, Institute of Biology, pp. 181–212.

Taylor, L. R., Taylor, R. A. J., Woiwod, I. P. and Perry, J. N. (1983) Behavioural dynamics. *Nature* **303**:801–804.

Taylor, L. R., Woiwod, I. P. and Perry, J. N. (1978) The density-dependence of spatial behaviour and the rarity of randomness. *J. anim. Ecol.* **47**:383–406.

Taylor, L. R., Woiwod, I. P. and Perry, J. N. (1980) Variance and the large scale spatial stability of aphids, moths and birds. *J. anim. Ecol.* **49**:831–854.

Taylor, L. R., Woiwod, I. P. and Taylor, R. A. J. (1979) The migratory ambit of the hop aphid and its significance in aphid population dynamics. *J. anim. Ecol.* **48**:955–972.

Taylor, R. A. J. and Taylor, L. R. (1979) 'A behavioural model for the evolution of spatial dynamics', in *Population Dynamics*, eds. R. M. Anderson, D. D. Turner and L. R. Taylor, Blackwell, Oxford, pp. 1–27.

Thornback, N. (1983) *The factors determining the abundance of* Metopolophium dirhodum *(Walk.), the rose grain aphid.* Ph.D thesis, University of East Anglia.

Thurston, R. (1970) Toxicity of trichome exudates of *Nicotiana* and *Petunia* species to tobacco hornworm larvae. *J. Econ. Ent.* **63**:272–274.

Thurston, R., Smith, W. T. and Cooper, B. P. (1966) Alkaloid secretion by trichomes of *Nicotiana* species and resistance of aphids. *Entomol. exp. appl.* **9**: 428–432.

Tingey, W. M. and Laubengayer, J. E. (1981) Defense against the green peach aphid and potato leafhopper by glandular trichomes of *Solanum berthaultii. J. Econ. Ent.* **74**:721–25.

Tingey, W. M and Sinden, S. L. (1982) Glandular pubescence, glycoalkaloid composition, and resistance to the green peach aphid, potato leaf hopper, and potato flea beetle in *Solanum berthaultii. Amer. Potato J.* **59**:95–106.

Tjallingii, W. F. (1978) Mechanoreceptors of the aphid labium. *Entomol. exp. appl.* **24**:531–537.

Toba, H. H., Paschke, J. D. and Friedman, S. (1967) Crowding as the primary factor in the production of the agamic alate form of *Therioaphis maculata* (Homoptera: Aphididae). *J. Insect Physiol* **13**:381–396.

Tomiuk, J. and Wöhrmann, K. (1980) Enzyme variability in populations of aphids. *Theor. appl. Genet.* **57**:125–127.

Tomiuk, J. and Wöhrmann, K. (1981) Changes of the genotype frequencies at the MDH-Locus in populations of *Macrosiphum rosae* (L.) (Hemiptera, Aphididae). *Biol. Zbl.* **100**:631–640.

Tomiuk, J. and Wöhrmann, K. (1982) Comments on the genetic stability of aphid clones. *Experientia* **38**:320–321.

Toth, L. (1940) The protein metabolism of the aphids. *Ann. Mus. Hist-nat. Hungar. Pars. Zool.* **33**:167–171.

Trager, W. (1970) *Symbiosis.* Van Nostrand, New York etc.

Ueda, N. and Takada, H. (1977) Differential relative abundance of green-yellow and red forms of *Myzus persicae* (Sulzer) (Homoptera Aphididae) according to host plant and season. *Appl. ent. Zool.* **12**:124–133.

van Emden, H. F. (1972) 'Aphids as phytochemists', in *Phytochemical Ecology*, ed. J. B. Harborne, Academic Press, London. pp. 25–43.

van Emden, H. F. (1978) Insects and secondary plant substances—an alternative viewpoint with special reference to aphids, in *Biochemical Aspects of Plant and Animal Coevolution*, ed. J. B. Harborne. Academic Press, London. pp. 309–323.

van Emden, H. F. and Bashford, M. A. (1971) The performance of *Brevicoryne brassicae* and *Myzus persicae* in relation to plant age and leaf amino-acids. *Entomol. exp. appl.* **14**:349–360.

van den Bosch, R. and Messenger, P. S. (1973) *Biological Control.* Intertext, London.

van den Bosch, R., Hom, R., Matteson, P., Frazer, B. D., Messenger, P. S. and Davis, C. J. (1979) Biological control of the walnut aphid in California: Impact of the parasite *Trioxys pallidus. Hilgardia* **47**(1): 1–13.

Varley, G. C. (1967) 'The effects of grazing by animals on plant productivity', in *Secondary Productivity of Terrestrial Ecosystems* Vol. 2, ed. K. Petrusewicz, Warsaw, pp. 773–778.

Vereijken, B. H. (1979) *Feeding and Multiplication of Three Cereal Aphid Species and their Effect on Yield of Winter Wheat.* Agric. Res. Report 888, Wageningen.

Walters, K. F. A. and Dixon, A. F. G. (1982) Effect of host quality and crowding on the settling and take-off of cereal aphids. *Ann. appl. Biol.* **101**:211–218.

Walters, K. F. A. and Dixon, A. F. G. (1983) Migratory urge and reproductive investment in aphids: variation within clones. *Oecologia (Berl.)* **58**:70–75.

Walters, K. F. A. and Dixon, A. F. G. (1984) The effect of temperature and wind on the flight activity of cereal aphids. *Ann. appl. Biol.* **104**:17–26.

Walters, K. F. A. and Dixon, A. F. G. (1985) (in press).

Ward, S. A. and Dixon, A. F. G. (1982) Selective resorption of aphid embryos and habitat changes relative to life-span. *J. anim. Ecol.* **51**:859–864.

Ward, S. A. and Dixon, A. F. G. (1984) Spreading the risk, and the evolution of mixed strategies: seasonal variation in aphid reproductive biology. *Adv. Invert. Reprod.* 3:367–386.

Ward, S. A., Dixon, A. F. G. and Wellings, P. W. (1983*a*) The relation between fecundity and reproductive investment in aphids. *J. anim. Ecol.* **52**:451–461.

Ward, S. A., Leather, S. R. and Dixon, A. F. G. (1984) Temperature prediction and the timing of sex in aphids. *Oecologia* (*Berl.*) **62**:230–233.

Ward, S. A., Wellings, P. W. and Dixon, A. F. G. (1983*b*) The effect of reproductive investment on pre-reproductive mortality in aphids. *J. anim. Ecol.* **52**:305–313.

Watson, M. A., Heathcote, G. D., Lauckner, F. B. and Sowray, P. A. (1975) The use of weather data and counts of aphids in the field to predict the incidence of yellowing viruses of sugar-beet crops in England in relation to the use of insecticides. *Ann. appl. Biol.* **81**: 181–198.

Watson, S. J. (1983) *Effects of weather on the numbers of cereal aphids.* Ph.D. thesis, University of East Anglia.

Watt, A. D. and Dixon, A. F. G. (1981) The role of cereal growth stages and crowding in the induction of alatae in *Sitobion avenae* and its consequences for population growth. *Ecol. Ent.* **6**:441–447.

Way, M. J. (1963) Mutualism between ants and honeydew-producing Homoptera. *Ann. Rev. Ent.* **8**:307–344.

Way, M. J. (1967) The nature and causes of annual fluctuations in numbers of *Aphis fabae* Scop. on field beans (*Vicia faba*). *Ann. appl. Biol.* **59**:175–188.

Way, M. J. (1968) 'Intra-specific mechanisms with special reference to aphid populations', in *Insect Abundance*, ed. T. R. E. Southwood, Blackwell, Oxford, pp. 18–36.

Way, M. J. (1971) A prospect of pest control. Inaugural Lecture, Imp. Coll. Sci. Technology, London, pp. 127–162.

Way, M. J. and Banks, C. J. (1964) Natural mortality of eggs of the black bean aphid *Aphis fabae* Scop., on the spindle tree, *Euonymus europaeus* L. *Ann. appl. Biol.* **54**: 255–267.

Way, M. J. and Banks, C. J. (1967) Intraspecific mechanisms in relation to the natural regulation of numbers of *Aphis fabae* Scop. *Ann. appl. Biol.* **59**:189–205.

Way, M. J. and Banks, C. J. (1968) Population studies on the active stages of the black bean aphid *Aphis fabae* Scop., on its winter host *Euonymus europaeus* L. *Ann. appl. Biol.* **62**: 177–197.

Way, M. J. and Cammell, M. (1970) 'Aggregation behaviour in relation to food utilization by aphids', in *Animal Populations in Relation to their Food Resources*, ed. A. Watson, Blackwell, Oxford, pp. 229–247.

Way, M. J., and Cammell, M. E. (1973) 'The problem of pest and disease forecasting—possibilities and limitations as exemplified by work on the bean aphid, *Aphis fabae*', in *Proc. 7th Br. Insecticides and Fungicides Conf.*, pp. 933–954.

Węgorek, W. and Krzymańska, J. (1971) Further studies on the resistance of lupine to the pea aphid (*A. pisum* Harris). *Prace Nauk. Inst. Ochrony Roślin* **13**:7–23.

Weismann, L. (1967) Die Populationsdynamik der schwarzen Rübenblattlaus *Aphis fabae* Scop. an der Zuckerrube als Grundlage der Schadensprognose. *Z. angew. Entomol.* **59**:1–15.

Wellings, P. W. (1981) The effect of temperature on the growth and reproduction of two closely related aphid species on sycamore. *Ecol. Ent.* **6**:209–214.

Wellings, P. W., Chambers, R. R. and Dixon, A. F. G. (1985) Sycamore aphid numbers and population density. I. Some Patterns. (in press).

Wellings, P. W. and Dixon, A. F. G. (1983) Physiological constraints on the reproductive activity of the sycamore aphid: the effect of developmental experience. *Entomol. exp. appl.* **34**:227–232.

Wellings, P. W., Leather, S. R. and Dixon, A. F. G. (1980) Seasonal variation in reproductive potential: a programmed feature of aphid life cycles. *J. anim. Ecol.* **49**:975–985.

Wellington, W. S. (1983) Biometerology of dispersal. *Bull. ent. Soc. Amer.* **29**:24–29.

Wensler, R. J. D. (1962) Mode of host selection by an aphid. *Nature* **195**:830–831.

Wensler, R. J. and Filshie, B. K. (1969) Gustatory sense organs in the food canal of aphids. *J. Morph.* **129**:473–492.

White, P. L. (1970) *The effect of aphids on tree growth.* Ph.D. thesis, University of Glasgow.

Whitham, T. G. (1978) Habitat selection by *Pemphigus* aphids in response to resource limitation and competition. *Ecology* **59**:1164–1176.

Whitham, T. G. (1979) Territorial behaviour of *Pemphigus* gall aphids. *Nature* **279**:324–325.

Whitham, T. G. (1980) The theory of habitat selection: examined and extended using *Pemphigus* aphids. *Amer. Nat.* **115**:449–466.

Williams, C. T. (1980) Low temperature mortality of cereal aphids. *I.O.B.C./W.P.R.S. Bull.* 111/4:63–66.

Williams, C. G. (1966) *Adaptation and Natural Selection.* Princeton University Press, Princeton.

Williams, G. C. (1975) *Sex and Evolution.* Princeton University Press, Princeton.

Williams, W. G., Kennedy, C. G., Yamamoto, R. T., Thacker, J. D. and Bordner, J. (198). 2-Tridecanone: a naturally occurring insecticide from the wild tomato *Lycopersicon hirsutum f. glabratum. Science* **207**:888.

Wink, M., Hartmann, T. Witte, L. and Rheinheimer, J. (1982) Interrelationship between quinolizidine alkaloid producing legumes and infesting insects: Exploitation of the alkaloid-containing phloem sap of *Cytisus scoparius* by the broom aphid *Aphis cytisorum. Z. Naturforsch.* **37**:1081–1086.

Wratten, S. D. (1977) Reproductive strategy of winged and wingless morphs of the aphids *Sitobion avenae* and *Metopolophium dirhodum. Ann. appl. Biol.* **85**:319–331.

Yagamuchi, H. (1976) Biological studies on the todo-fir aphid *Cinàra todicola* Inouye with special reference to its population dynamics and morph determination. *Bull. Gov. For. Exp. Stn.* **283**:1–102.

Young, J. P. W. (1981) Sib competition can favour sex in two ways. *J. theoret. Biol.* **88**:755–756.

Zeigler, H. (1975) 'Nature of transported substances', in *Transport in Plants. I. Phloem Transport*, eds. M. H. Zimmerman and J. A. Milburn, Springer, Berlin etc., pp. 59–100.

Zucker, W. V. (1982) How aphids choose leaves: the roles of phenolics in host selection by a galling aphid. *Ecology* **63**:972–981.

Index